AF361930

Electrocatalysts for Oxygen/Hydrogen-Involved Reactions

Electrocatalysts for Oxygen/Hydrogen-Involved Reactions

Editors

Jingqi Guan
Yin Wang

MDPI • Basel • Beijing • Wuhan • Barcelona • Belgrade • Manchester • Tokyo • Cluj • Tianjin

Editors
Jingqi Guan
College of Chemistry
Jilin University
Changchun
China

Yin Wang
College of Chemistry and
Materials Science
Inner Mongolia University for
Nationalities
Tongliao
China

Editorial Office
MDPI
St. Alban-Anlage 66
4052 Basel, Switzerland

This is a reprint of articles from the Special Issue published online in the open access journal *Molecules* (ISSN 1420-3049) (available at: www.mdpi.com/journal/molecules/special_issues/electrocatalysts_oxygenhydrogen_involved_reactions).

For citation purposes, cite each article independently as indicated on the article page online and as indicated below:

LastName, A.A.; LastName, B.B.; LastName, C.C. Article Title. *Journal Name* **Year**, *Volume Number*, Page Range.

ISBN 978-3-0365-4026-9 (Hbk)
ISBN 978-3-0365-4025-2 (PDF)

Contents

About the Editors

Jingqi Guan

Jingqi Guan received his B.S. degree and Ph.D. degree from Jilin University in 2002 and 2007. He was promoted to professor of chemistry in Jilin University in 2021. He worked as a postdoctoral research fellow in the University of California at Berkeley from 2012 to 2013 and in the Dalian Institute of Chemical Physics from 2014 to 2018. He has published more than 170 papers. His current research interests focus on the synthesis of mononuclear-multinuclear catalysts for energy storage and conversion.

Yin Wang

Yin Wang is working as a Distinguished Researcher in Inner Mongolia Minzu University, China. He received his Ph.D. in College of Chemistry and Chemical Engineering from BeInstitute of Technology in 2018, under the guidance of Professor Changwen Hu. His current research interests lie in the synthesis and applications of inorganic nanoscale materials as electrocatalyst for OER, HER, NRR, ORR, and 2eORR.

Preface to "Electrocatalysts for Oxygen/Hydrogen-Involved Reactions"

This Special Issue of {Molecules} presents some recent exciting developments in catalysis and energy chemistry. It covers a range of topics including important research on novel approaches to the construction of heterogeneous electrocatalysts for the hydrogen evolution reaction, oxygen evolution reaction, and oxygen reduction reaction, and the theoretical understanding of the reaction mechanisms. The Special Issue should prove to be a valuable resource for electrochemical scientists and postgraduate students seeking updated and critically important information in catalysis and energy chemistry. The articles are written by authorities in the field and are mainly focused on the synthesis, characterization, and structure–function relationship of oxygen/hydrogen-involved electrocatalysts, and the applications in water splitting, proton exchange membrane fuel cells and lithium-ion batteries. We hope that the readers will find these research articles valuable and thought provoking so that they may trigger further research in the quest for new developments in the field.

We are grateful to MDPI academic open access publishing for the timely efforts made by the editorial personnel, especially Ms. Marlene Zhang (Section Managing Editor).

Jingqi Guan and Yin Wang
Editors

Editorial

Electrocatalysts for Oxygen/Hydrogen-Involved Reactions

Jingqi Guan [1],* and Yin Wang [2],*

[1] Institute of Physical Chemistry, College of Chemistry, Jilin University, 2519 Jiefang Road, Changchun 130021, China
[2] Inner Mongolia Key Laboratory of Carbon Nanomaterials, Nano Innovation Institute (NII), College of Chemistry and Materials Science, Inner Mongolia Minzu University, Tongliao 028000, China
* Correspondence: guanjq@jlu.edu.cn (J.G.); ywang@imun.edu.cn (Y.W.)

Citation: Guan, J.; Wang, Y. Electrocatalysts for Oxygen/Hydrogen-Involved Reactions. *Molecules* **2022**, *27*, 2628. https://doi.org/10.3390/molecules27092628

Received: 13 April 2022
Accepted: 18 April 2022
Published: 19 April 2022

Publisher's Note: MDPI stays neutral with regard to jurisdictional claims in published maps and institutional affiliations.

Oxygen/hydrogen-involved reactions are key reactions in many energy-related technologies, such as electrolytic water, electrocatalytic carbon dioxide reduction, electrochemical ammonia synthesis, rechargeable metal–air batteries, and renewable fuel cells. The development of highly active and stable hydrogen evolution reaction (HER), oxygen evolution reaction (OER), and oxygen reduction reaction (ORR) electrocatalysts is key to the large-scale implementation of these technologies. In addition, novel, high-performance HER, OER, and ORR catalysts could be predicted by theoretical investigations. This Special Issue aims to provide a broad survey of the most recent advances in oxygen/hydrogen-involved reactions and the applications in proton exchange membrane fuel cells (PEMFCs) and lithium-ion batteries (LIBs).

In this Special Issue, nine original research articles containing some recent experimental and theoretical advances in HER, OER, and ORR are reported. Five articles deal with the design and application of electrocatalysts for oxygen and hydrogen evolution reactions. Zhou et al. reported a 2D SnSe film in situ grown on a mica substrate using the molecular beam epitaxy method. The defective Sn structure was modulated by controlling experiment conditions and a p-type semiconductor of SnSe was formed. First-principles calculations revealed that Sn vacancies can serve as reactive centers and promote electron migration for HER [1]. Song et al. synthesized PtNi alloy loaded on zeolite imidazolate framework (ZIF), derived from porous nitrogen-doped carbon, and found that the optimized PtNi/MNC-1-6 catalyst exhibited a comparable electrocatalytic HER activity with that of commercial 20 wt% Pt/C and showed a quite small Tafel slope of 21.5 mV dec^{-1} [2]. Sondermann et al. reported a series of NiM-BTC (M = Co, Fe, BTC = 1,3,5-benzanetricarboxylate) electrocatalysts for alkaline OER. They found that $Ni_{10}Co$-BTC and $Ni_{10}Co$-BTC/KB composites underwent structural changes at high potentials, and the generated $Ni(OH)_2$ contributed to the high electrocatalytic performance and stability of the OER [3]. Liu et al. developed a novel $Sm_{0.5}Sr_{0.5}Co_{0.8}Ni_{0.2}O_{3-\delta}$ (SSCN82) nanofiber structure compound as a bifunctional electrocatalyst for OER and ORR, which afforded an onset potential of 1.39 V for OER and a Tafel slope of 111.8 mV dec^{-1} for ORR [4]. Woitasske et al. reported a facile microwave-induced heating method to prepare Pt nanoparticles on reduced graphite oxide (Pt-NP@rGO). HRTEM revealed that 2~6 nm Pt particles were homogeneously distributed on the rGO without aggregation, which showed superior catalytic activity for the ORR [5]. Two articles deal with computational studies of HER, ORR, and OER electrocatalysts. Cherif et al. used density functional theory (DFT) to study the effects of fluorine modification of FeN_x-doped and N-doped carbon catalysts on the ORR performance for PEM fuel cells. They studied 12 possible FeN_x-doped and N-doped carbon models in the absence or presence of fluorine atoms. The results showed that metal-free catalytic N-sites were difficult to combine with F atoms, and in FeN_x-doped carbon configuration, only one F atom coordinated with the FeN_x moiety helped to improve its ORR activity [6]. Yang et al. calculated the electrocatalytic thermodynamics of HER, OER, and ORR on three two-dimensional, Fe-based, metal–organic frameworks (2D Fe-MOFs). They investigated

NH, O, and S ligands influencing the electronic structure and catalytic performance of 2D-Fe-MOFs and confirmed that Fe-O MOF displayed an optimal $\triangle G_H$ of 0.008 eV for HER, Fe-NH MOF exhibited the lowest η_{ORR} of 0.38 V for ORR, and three Fe-MOFs possessed poor OER performance ($\eta_{OER} > 0.92$ V) [7]. Two articles study the design of electrodes for use in fuel cells and batteries. Song et al. investigated the degradation of Pt electrocatalyst in PEMFCs under conditions of high energy efficiency. The corrosion of carbon support and the agglomeration of Pt nanoparticles under high operation potential (0.8 V) and high current density (1000 mA cm^{-2}) led to the cell voltage declining more than 13 % after the operation for 64 h [8]. Wang et al. reported a nanosheet-assembled SnO$_2$-integrated anode for LIBs on which a large reversible specific capacity of 637.2 mAh g^{-1} could be achieved. Such excellent capacity was a benefit of the rational design based on structural engineering to boost the synergistic effects of the integrated electrode [9].

In conclusion, oxygen/hydrogen-involved reactions are undoubtedly a hot and crucial topic in energy-related fields. Several research groups have contributed to the advancement of this topic by designing novel heterogeneous HER, OER, and ORR electrocatalysts, characterizing their structure, evaluating the catalytic performance, and studying the structure–activity relationship. Herein, important advancements in oxygen/hydrogen-involved reactions are revealed. We thank all of the authors for their valuable contributions to this Special Issue, all the peer reviewers for their valuable comments, criticisms, and suggestions, and the staff members of MDPI for the editorial support.

Conflicts of Interest: The authors declare no conflict of interest.

References

1. Zhou, Q.; Wang, M.; Li, Y.; Liu, Y.; Chen, Y.; Wu, Q.; Wang, S. Fabrication of Highly Textured 2D SnSe Layers with Tunable Electronic Properties for Hydrogen Evolution. *Molecules* **2021**, *26*, 3319. [CrossRef] [PubMed]
2. Song, X.; He, Y.; Wang, B.; Peng, S.; Tong, L.; Liu, Q.; Yu, J.; Tang, H. PtNi Alloy Coated in Porous Nitrogen-Doped Carbon as Highly Efficient Catalysts for Hydrogen Evolution Reactions. *Molecules* **2022**, *27*, 499. [CrossRef] [PubMed]
3. Sondermann, L.; Jiang, W.; Shviro, M.; Spiess, A.; Woschko, D.; Rademacher, L.; Janiak, C. Nickel-Based Metal-Organic Frameworks as Electrocatalysts for the Oxygen Evolution Reaction (OER). *Molecules* **2022**, *27*, 1241. [CrossRef] [PubMed]
4. Liu, X.; Wang, Y.; Fan, L.; Zhang, W.; Cao, W.; Han, X.; Liu, X.; Jia, H. Sm$_{0.5}$Sr$_{0.5}$Co$_{1-x}$Ni$_x$O$_3$-delta—A Novel Bifunctional Electrocatalyst for Oxygen Reduction/Evolution Reactions. *Molecules* **2022**, *27*, 1263. [CrossRef] [PubMed]
5. Woitassek, D.; Lerch, S.; Jiang, W.; Shviro, M.; Roitsch, S.; Strassner, T.; Janiak, C. The Facile Deposition of Pt Nanoparticles on Reduced Graphite Oxide in Tunable Aryl Alkyl Ionic Liquids for ORR Catalysts. *Molecules* **2022**, *27*, 1018. [CrossRef] [PubMed]
6. Cherif, M.; Dodelet, J.P.; Zhang, G.; Glibin, V.P.; Sun, S.; Vidal, F. Non-PGM Electrocatalysts for PEM Fuel Cells: A DFT Study on the Effects of Fluorination of FeNx-Doped and N-Doped Carbon Catalysts. *Molecules* **2021**, *26*, 7370. [CrossRef]
7. Yang, X.; Feng, Z.; Guo, Z. Theoretical Investigation on the Hydrogen Evolution, Oxygen Evolution, and Oxygen Reduction Reactions Performances of Two-Dimensional Metal-Organic Frameworks Fe$_3$(C$_2$X)$_{12}$ (X = NH, O, S). *Molecules* **2022**, *27*, 1528. [CrossRef]
8. Song, J.; Ye, Q.; Wang, K.; Guo, Z.; Dou, M. Degradation Investigation of Electrocatalyst in Proton Exchange Membrane Fuel Cell at a High Energy Efficiency. *Molecules* **2021**, *26*, 3932. [CrossRef]
9. Wang, X.; Zhao, X.; Wang, Y. A Nanosheet-Assembled SnO$_2$-Integrated Anode. *Molecules* **2021**, *26*, 6108. [CrossRef] [PubMed]

Article

Theoretical Investigation on the Hydrogen Evolution, Oxygen Evolution, and Oxygen Reduction Reactions Performances of Two-Dimensional Metal-Organic Frameworks $Fe_3(C_2X)_{12}$ (X = NH, O, S)

Xiaohang Yang [1], Zhen Feng [2,3,*] and Zhanyong Guo [2]

[1] School of Science, Henan Institute of Technology, Xinxiang 453000, China; yangxh@hait.edu.cn
[2] School of Materials Science and Engineering, Henan Institute of Technology, Xinxiang 453000, China; guozhanyong123@126.com
[3] School of Physics, Henan Normal University, Xinxiang 453007, China
* Correspondence: fengzhen@hait.edu.cn

Abstract: Two-dimensional metal-organic frameworks (2D MOFs) inherently consisting of metal entities and ligands are promising single-atom catalysts (SACs) for electrocatalytic chemical reactions. Three 2D Fe-MOFs with NH, O, and S ligands were designed using density functional theory calculations, and their feasibility as SACs for hydrogen evolution reaction (HER), oxygen evolution reaction (OER), and oxygen reduction reaction (ORR) was investigated. The NH, O, and S ligands can be used to control electronic structures and catalysis performance in 2D Fe-MOF monolayers by tuning charge redistribution. The results confirm the Sabatier principle, which states that an ideal catalyst should provide reasonable adsorption energies for all reaction species. The 2D Fe-MOF nanomaterials may render highly-efficient HER, OER, and ORR by tuning the ligands. Therefore, we believe that this study will serve as a guide for developing of 2D MOF-based SACs for water splitting, fuel cells, and metal-air batteries.

Keywords: two-dimensional metal-organic framework; ligand; single-atom catalysts; hydrogen evolution reaction; oxygen evolution reaction; oxygen reduction reaction

Citation: Yang, X.; Feng, Z.; Guo, Z. Theoretical Investigation on the Hydrogen Evolution, Oxygen Evolution, and Oxygen Reduction Reactions Performances of Two-Dimensional Metal-Organic Frameworks $Fe_3(C_2X)_{12}$ (X = NH, O, S). *Molecules* **2022**, *27*, 1528. https://doi.org/10.3390/molecules27051528

Academic Editors: Jingqi Guan and Yin Wang

Received: 30 January 2022
Accepted: 22 February 2022
Published: 24 February 2022

Publisher's Note: MDPI stays neutral with regard to jurisdictional claims in published maps and institutional affiliations.

1. Introduction

The greenhouse effect, air pollution, ozone depletion, and fossil fuel depletion are all major challenges for our society's progress in the 21st century [1]. To address the environmental deterioration and energy challenges, it has become a major priority to increase the research and development of low-cost, efficient, and renewable energy storage and conversion devices, such as fuel cells, metal-air cells, and water decomposition [2]. At the United Nations Climate Summit 66 countries pledged to achieve net-zero carbon emissions by 2050 [3]. A promising energy conversion technology is the unitized regenerative fuel cell. It works like a fuel cell and inversely as a water electrolyzer to produce H_2 and O_2 to feed the fuel cell. Hence, multifunctional electrocatalysts play key roles [4]. However, because of the high overpotential, low activity, and poor selectivity, it is extremely desirable to develop sustainable and low-cost functional electrode materials with high energy density, excellent rate capability, and good cycling stability [5].

Since an isolated Pt single atom anchored in FeO_x showed remarkable catalytic performance for CO oxidation [6], single-atom catalysts (SACs) have been considered next-generation electrode candidates. SACs contain isolated single-metal atoms dispersed on a support surface, and represent the ultimate limit of atom use efficiency for catalysis [7]. Some experimental and computational studies show that SACs are promising for precise control of catalytic reactions, such as the hydrogen evolution reaction (HER) [8,9], the oxygen evolution reaction (OER) [10,11], the oxygen reduction reaction (ORR) [12], the nitrogen reduction reaction (NRR) [13], the carbon dioxide reduction reaction (CO_2RR) [14], and CO/NO oxidation [15].

Unlike typical SACs, two-dimensional metal-organic frameworks (2D MOF) contain metal entities and organic ligands, indicating that they could be used as SACs [16]. MOF monolayers have highly exposed metal atoms, uniformly dispersed against agglomeration [16]. Similar to SACs, the metal entities in MOFs could effectively modify charge redistribution and boost chemical reactions [17].

Recently, more and more 2D MOF sheets have been experimentally synthesized and theoretically predicted for use as catalysts [18]. The $Cu_3(C_6S_6)$ MOF cathode enables a high reversible capacity for lithium-ion batteries [19]. The $Rh_3C_{12}S_{12}$ MOF exhibits a low limiting potential of -0.43 V for CO_2RR [20]. $Mo_3C_{12}N_{12}H_{12}$ MOF exhibits a low overpotential of 0.18 V for NRR [21]. $Mo_3(C_2O)_{12}$ MOF could achieve a low limiting potential of -0.36 V for NRR via the distal pathway [22]. These MOFs could also be used as electrocatalysts for ORR [23], OER [24], and HER [25].

Due to the above results, the NH, O, and S ligands were adopted to design 2D Fe-MOFs. The potential as SACs for the HER, OER, and ORR was systematically explored. The Fe-MOF exhibits atomically thin like 2D graphene, and they display excellent structural stability. The NH, O, and S ligands could tune charge redistribution in Fe-MOF and catalysis performance. The Fe-O MOF displays $\Delta G_H = 0.08$ eV for HER, Fe-NH MOF exhibits $\eta^{ORR} = 0.38$ V for ORR, and they possess poor OER catalysis performance ($\eta^{OER} > 0.92$ V). Our work highlights the effect of ligands, and could guide the development of highly effective SACs based on 2D MOFs.

2. Results and Discussion

2.1. Geometry and Stability

The unit cells of the studied 2D Fe-MOF monolayers are depicted in Figure 1. There are three types of ligating atoms between Fe atoms and graphene nanosheets (Figure S1). Therefore, different symbols were denoted by different ligating atoms: Fe-NH-MOF for NH ligating atoms, Fe-O-MOF for O ligating atoms, and Fe-S-MOF for S ligating atoms, respectively. These 2D MOF sheets are also atomically thin like grapheme, but each unit cell consists of 3 Fe atoms, 24 carbon atoms, and 12 ligating atoms.

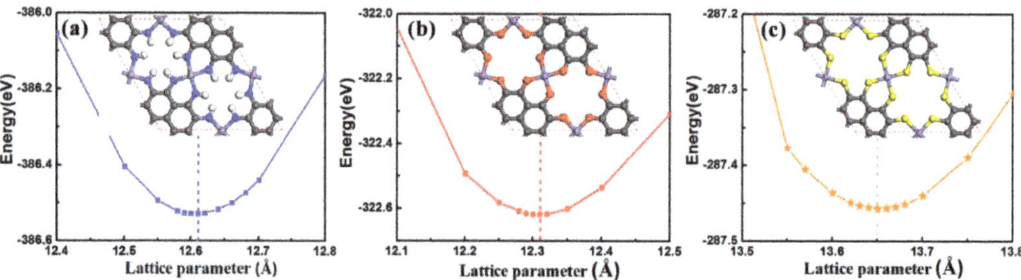

Figure 1. Optimized lattice parameters of (**a**) Fe-NH-MOF, (**b**) Fe-O-MOF, and (**c**) Fe-S-MOF monolayers.

To optimize the atomic structures of these three Fe-MOFs, the variations of energies vs. the lattice constants are also shown in Figure 1, their lattice constants are optimized to be 12.61 Å for Fe-NH MOF, 12.31 Å for Fe-O MOF, and 13.65 Å for Fe-S MOF, respectively (Table 1 and Table S1). These optimized lattice constants agree with previous investigations [22,26,27]. The bond lengths of Fe-N, Fe-O, and Fe-S are 1.85 Å, 1.83 Å, and 2.15 Å, respectively, and the bond lengths of C-N, C-O, and C-S are 1.35 Å, 1.30 Å, and 1.74 Å, respectively, due to differences in the atomic radius of N (r = 70 pm), O (r = 66 pm), and S (r = 104 pm). The diameters of the holes in Fe-MOF sheets vary as well by 3.44 for Fe-NH MOF, 5.17 Å for Fe-O MOF, and 5.74 Å for Fe-S MOF, respectively (Table 1).

Table 1. Calculated geometric parameters (lattice constants (l_a), the bond length of Fe-N (O, S) ($D_{\text{Fe-N (O,S)}}$), C-N (O,S) ($D_{\text{C-N (O,S)}}$), and diameter of the hole (Φ) (Å)), total magnetic moment (M_{tot}, μ_B) and Bader charge (Q_{Fe} (e), $Q_{\text{N,O,S}}$ (e)) of Fe-NH-MOF, Fe-O-MOF, and Fe-S-MOF monolayers.

Materials	l_a (Å)	$D_{\text{Fe-N (O,S)}}$ (Å)	$D_{\text{C-N (O,S)}}$ (Å)	Φ (Å)	M_{tot} (μ_B)	Q_{Fe} (e)	$Q_{\text{N,O,S}}$ (e)
Fe-NH-MOF	12.61	1.85	1.35	3.44	6.00	−1.26	+0.83
Fe-O-MOF	12.31	1.83	1.30	5.17	10.45	−1.51	+1.07
Fe-S-MOF	13.65	2.15	1.74	5.74	9.25	−0.58	+0.13

We examined the stability of three Fe-MOF monolayers after attaining their unique structures, because the good stability of the given materials is a prerequisite for their practical uses. Notably, high-quality 2D Fe-S MOF have been experimentally synthesized [27], however, their thermal stabilities were still performed using first-principles plus ab initio molecular dynamics simulations (AIMD). At a temperature of 500 K, a total time process of 3000 fs with a time step of 1 fs was implemented using specified 2 × 1 × 1 rectangular supercells (included 102 atoms for Fe-NH MOF and 78 atoms for Fe-O/ Fe-S MOFs). The variations of total energy and temperature and the final snapshots in 3000 fs are depicted in Figure 2. Their total energies and temperature exhibit up and down trends within a fixed range. These final structures display lack of structural distortion and no bond-breaking. It slight up and down changes in final plane structure can be seen. These results revealed that these three Fe-MOF monolayers could maintain their original atomic structures at a high temperature of 500 K, implying their exceptional thermal stability. The high stability may result from the large π-bonds of high-symmetric sp²-C atoms in graphene nanosheets, which agrees with previous work [22,27]. Notable are the slight up and down changes in the plane. The plane structure shows some fluctuation changes.

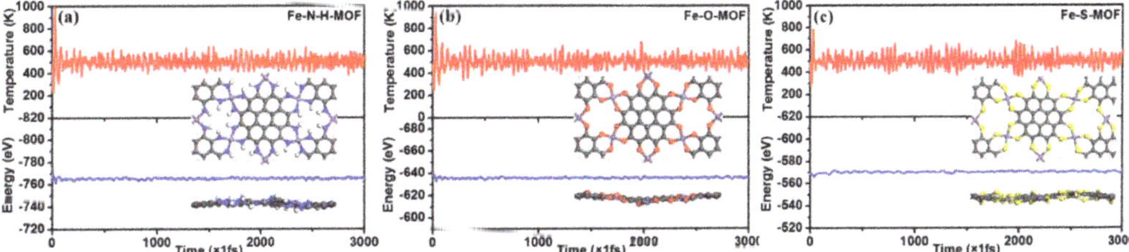

Figure 2. The AIMD simulations of (**a**) Fe-NH-MOF, (**b**) Fe-O-MOF, and (**c**) Fe-S-MOF monolayers at 500 K during the timescale of 3 ps.

We further perform Bader charge analysis to investigate the chemical bonding in these Fe-MOF monolayers. The electron localization function (ELF) map and isosurfaces of ELF with a value of 0.50 au are plotted in Figure 3a–c, the Fe loses 0.58–1.51 e, and the N, O, and S gain 0.13–1.07 e, which contributes to their robust ionic bonds. Note that the positive Fe atom may be used as an active site for chemical reactions.

2.2. Electronic Property

Previous research has shown that the electronic structures of 2D-based catalysts have a significant impact on their catalytic efficiency [28]. Thus, we computed their band structures and density of states (DOS) with the DFT + U method [29]. As shown in Figure 4a, the Fe-O MOF and Fe-S MOF display intrinsic metallicity due to the several bands at the Fermi level, while Fe-NH MOF is a semiconductor with band gaps of 0.56 eV for spin up and 0.89 eV for spin down. Thus, the high electrical conductivity of Fe-O and Fe-S MOFs should ensure rapid charge transfer in electrochemical reactions. It can be clearly observed that all three Fe-MOF nanomaterials possess spin splitting of band structures, producing magnetism. The computed total magnetic moment (M_{tot}) of the primitive cells of Fe-NH-MOF, Fe-O-

MOF, and Fe-S-MOF monolayers are 6.00 µB, 10.45 µB, and 9.25 µB, respectively. Further study of magnetism comes from the spin-charge density in Figure 3(a3,b3,c3), the spin-up densities are mainly around the Fe atoms, matching their total magnetic moments (Table 1).

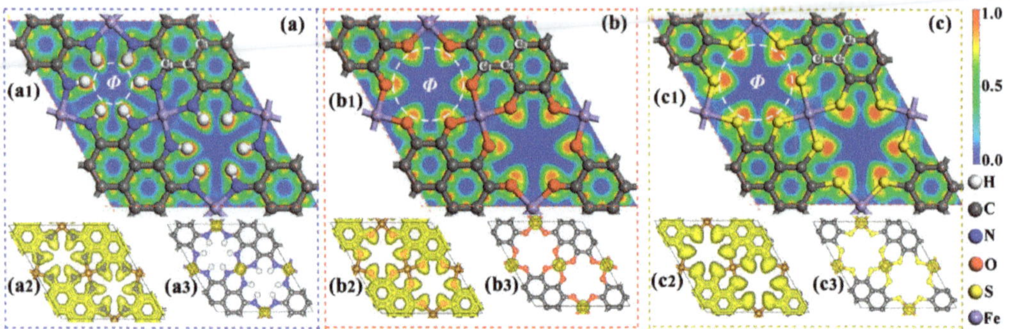

Figure 3. Geometries (1), electron localization function (ELF) (2), and spin density (3) of (**a**) Fe-NH-MOF, (**b**) Fe-O-MOF, and (**c**) Fe-S-MOF monolayers.

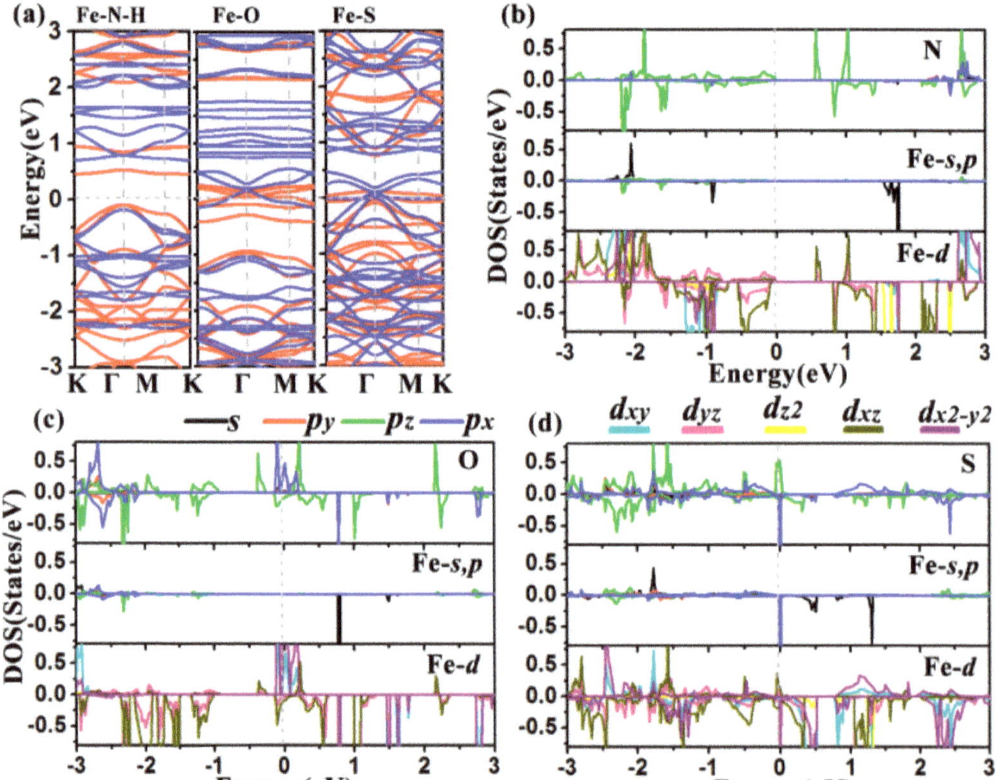

Figure 4. Band structures (**a**) and projected density of states (PDOS) of (**b**) Fe-N-H-MOF, (**c**) Fe-O-MOF, and (**d**) Fe-S-MOF monolayers. The Fermi levels (EF) are set to 0 eV.

Previous research suggested that metallic characteristics and strongly spin-polarized Fe atoms could enhance the chemical catalysis process [30]. The density of states (DOS) of these three Fe-MOFs was then calculated to further understand them better. The projected density of states (PDOS) of Fe, N, O, S, and C elements are further plotted in Figure 4b–d. There are obvious hybridizations between Cu-dyz and Cu-dxz orbitals and N-pz orbitals (Figure 4b), confirming the strong bond between Fe and N atoms. We also concluded the semiconductor character for Fe-NH MOF since there are no states at the Fermi level. For the Fe-O MOF in Figure 4c, the metallic feature mainly comes from the contributions of Fe-dxy, Fe-dx^2-y^2, O-px, and O-pz orbitals, which also mainly give the spin magnetism and hybridizations of the Fe-O bond. For Fe-S MOF in Figure 4d, S-px, S-pz, Fe-px and Fe-dx^2-y^2 existing at the Fermi level, these orbitals endow their metallicity, spin magnetism, and strong bond to Fe-S.

2.3. HER

The hydrogen adsorption free energies are estimated under various configurations in this part to assess the HER activity of these Fe-MOFs. Four representative adsorption sites are chosen to show the HER catalysis activity, which are Fe, N/O/S, C_1, C_2, and C_3 atoms in Figure 3. The corresponding adsorption structures are displayed in Figures S2–S4. The calculated HER free energy diagrams of three Fe-MOFs at a potential $U = 0$ relative to the standard hydrogen electrode at pH = 0 are plotted in Figure 5. For Fe-NH MOF in Figure 5a, the Gibbs free energies of hydrogen adsorption (ΔG_H) on Fe and N sites are 0.16 eV and 0.45 eV, while the ΔG_H is more than 0.88 eV on C sites, suggesting that the optimized HER activity is 0.16 eV. The ΔG_H of Fe-O MOF on Fe and O atoms are 0.60 eV and 0.08 eV, respectively, the C atoms also exhibit a poor HER catalysis performance and the $\Delta G_H > 0.67$ eV. Similarly, the Fe-S MOF monolayer possesses poor HER catalytic behavior due to the high hydrogen adsorption free energies ($\Delta G_H > 0.37$ eV).

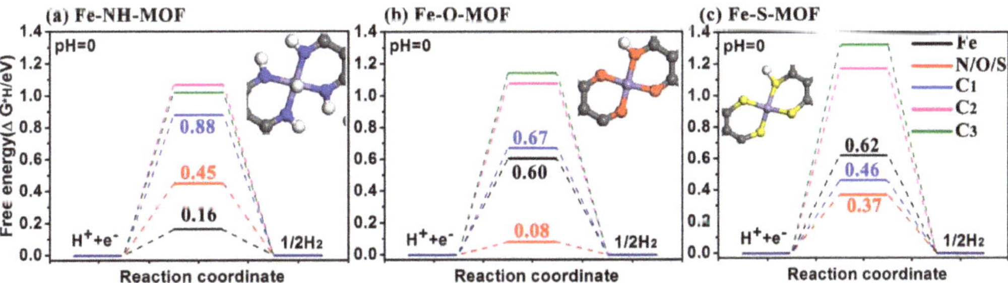

Figure 5. Calculated free energy diagram for hydrogen evolution on (**a**) Fe-NH-MOF, (**b**) Fe-O-MOF, and (**c**) Fe-S-MOF monolayers at a potential $U = 0$ relative to the standard hydrogen electrode at pH = 0. Insets are corresponding optimized adsorbed H intermediates.

The reason can be deduced from the Bader charge analysis in Table 1, the O gains more electrons (+1.07 e) showing higher catalysis activity, the S and N gain fewer electrons (+0.13–+0.83 e) displaying poor catalysis properties. In comparison with the findings of previous studies (Table 2), the Fe-O MOF displays the small or comparable hydrogen adsorption free energy ($\Delta G_H = 0.08$ eV), implying its excellent HER electrocatalytic activity.

Table 2. Comparison of the HER (ΔG_H, eV), OER (η^{OER}, V), and ORR (η^{ORR}, V) catalysis performance in our results with previous literatures.

Materials	ΔG_H (eV)	η^{OER} (V)	η^{ORR} (V)	Materials	ΔG_H (eV)	η^{OER} (V)	η^{ORR} (V)
Fe-NH-MOF	0.16	0.92	0.38	IrO$_2$ [31]	-	0.45–0.59	-
Fe-O-MOF	0.08	1.00	0.85	Co-BP [32]	-	0.42	0.36
Fe-S-MOF	0.37	1.22	0.75	Ni-BP [32]	-	0.44	0.29
V-W$_2$B$_2$O$_2$ [33]	0.01–0.15	-	-	Pt-BP [32]	-	0.25	0.32
ZnW$_2$B$_2$O$_2$ [33]	0.14–0.26	-	-	Fe-BHT [24]	-	0.88	-
Zn@InSe [34]	0.02	-	-	Ir$_3$(HITP)$_2$ [35]	-	-	0.31
Ni@PR-GDY [36]	−0.05	0.29	0.38	Rh$_3$(HITP)$_2$ [35]	-	-	0.37

2.4. OER

The OER electrocatalytic activity of these three Fe-MOF monolayers was then evaluated. According to previous work, the OER process should consist of four elementary steps [37]. They are (1) *+ H$_2$O $\rightarrow$ *OH+ H$^+$, (2) *OH $\rightarrow$ *O + H$^+$, (3) *O + H$_2$O $\rightarrow$ *OOH + H$^+$, and (4) *OOH $\rightarrow$ *+ O$_2$ + H$^+$, respectively. The optimized atomic structures of *OH, *O, and *OOH intermediates on three Fe-MOF monolayers are diagnosed by considering different adsorbed sites and conformations (Figures S2–S4). It is found that their reaction sites are the same as Fe atoms (Figure 6), which is in agreement with those of other MOF materials [22,24].

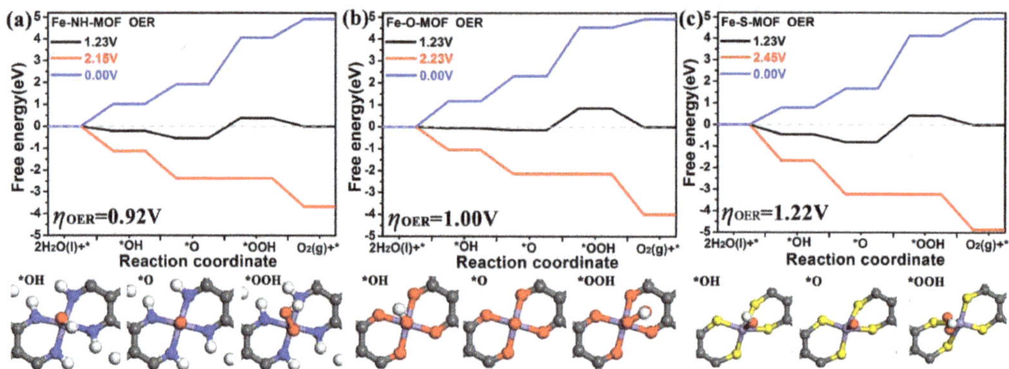

Figure 6. Calculated free energy diagram for OER on (**a**) Fe-NH-MOF, (**b**) Fe-O-MOF, and (**c**) Fe-S-MOF monolayers at a potential U = 0, 1.23, and working potentials relative to the standard hydrogen electrode at pH = 0. Insets are corresponding OER intermediates.

The OER free energy diagram at 0.00 V, 1.23 V, and work potential is depicted in Figure 6. The OER free energy diagram of Fe-NH MOF plotted in Figure 6a indicates that the third step possesses the biggest uphill, and the elementary step *O + H$_2$O $\rightarrow$ *OOH + H$^+$ is the rate-limiting step. When the electrode is 2.15 V, all four-element steps are downhill. Thus, the calculated OER overpotential (η^{OER}) is 0.92 V. The working potentials of Fe-O MOF and Fe-S MOF are 2.23 V and 2.45 V, where it can be deduced that their OER overpotentials are 1.00 V and 1.22 V. Therefore, the Fe-NH MOF exhibits the best OER catalytic activity for converting H$_2$O to O$_2$ in this study. We further compare the OER performance of Fe-MOF with the recent catalysts in Table 2. It is worth noting that the OER overpotential of Fe MOF (η^{OER} = 0.92–1.22 V) is 2–3 times that of the best-known OER catalyst IrO$_2$ (0.45–0.59 V) [31], implying their poor OER electrocatalytic activity. The other methods should be adopted to tune the OER electrocatalytic activity of 2D Fe MOF monolayer.

2.5. ORR

The ORR electrocatalytic activity of the Fe–MOF sheets is discussed in this section. Previous studies have shown that the O_2 dissociative pathway is difficult to achieve on 2D MOF materials, which are similar to Pt(111) [38] and several single-atom catalysts [28,36,39,40]. Here, the ORR can be regarded as the inverse process of OER, which is also described by four elementary steps (1) * + O_2 + H^+ → *OOH, (2) *OOH + H^+ → *O + H_2O, (3) *O + H^+ → *OH, and (4) *OH + H^+ → *+ H_2O, respectively. The optimized atomic structures of ORR intermediates on three Fe-MOF monolayers are the same as OER intermediates on them and the active sites are Fe atoms (Figures S2–S4).

The calculated ORR free energy diagrams of three Fe-MOFs are depicted in Figure 7. The blue, black, and red lines represent the electrode potentials at 0.00 V, 1.23 V, and working potential. As displayed in Figure 7a, the elementary steps of Fe-NH MOF at 0.00 V are downhill. When the electrode potential is up to 0.85 V (which is defined as working potential, U_{work}), the first step is spontaneous, thus the * + O_2+ H^+ → *OOH step is the rate-limiting step, which is defined as working potential. Thus, the corresponding ORR overpotential is 0.38 V, which can be calculated by the equation (η^{ORR} = 1.23–0.85). Similarly, it is worth noting from Figure 7b,c, that the rate-limiting steps on Fe-O MOF and Fe-S MOF are the first and the fourth steps. Their corresponding working potentials (U_{work}) are 0.38 V and 0.48 V for Fe-O MOF and Fe-S MOF, indicating that their ORR overpotentials (η^{ORR}) are 0.85 V and 0.75 V. Therefore, the Fe-NH MOF possesses the lowest overpotential, and the η^{ORR} is even lower than the best ORR catalyst of Pt (0.45 V) [38]. We also compare this η^{ORR} value with that of other excellent ORR catalysts (Table 2), which suggests that the Fe-NH MOF could boost ORR electrocatalysis performance.

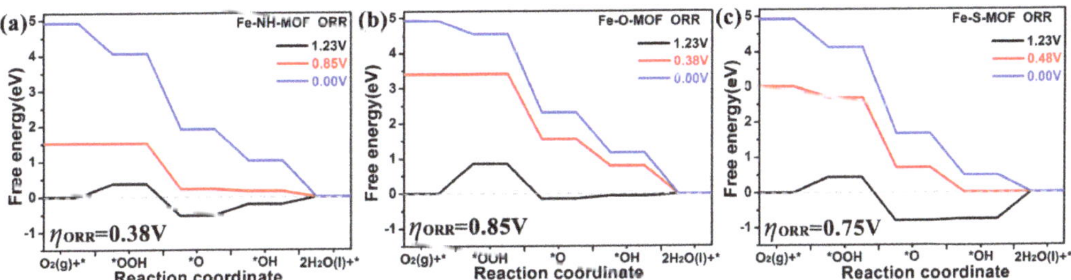

Figure 7. Calculated free energy diagram for hydrogen evolution on (**a**) Fe-NH-MOF, (**b**) Fe-O-MOF, and (**c**) Fe-S-MOF monolayers at a potential U = 0, 1.23, and working potentials relative to the standard hydrogen electrode at pH = 0.

To gain further insight into the catalysis property of Fe-MOF, we analyze the adsorption free energies of OOH, O, and OH species. Here we give the values of an ideal ORR catalyst, which are 3.69 eV for ΔG*OOH, 2.46 eV for ΔG*O, and 1.23 eV for ΔG*OH, respectively. Furthermore, the Sabatier principle claims that an ideal catalyst should serve moderate adsorbed energies for all reaction species. For Fe-NH MOF, the values are ΔG*OOH = 4.06 eV, ΔG*O = 1.92 eV, and ΔG*OH = 1.02 eV, the corresponding adsorbed energy differences are 0.10 eV (4.06–3.69), 0.54 eV (2.46–1.92), and 0.21 eV (1.23–1.02), respectively, which indicates that Fe-NH MOF exhibits strong adsorption to *O species. We also conclude that the adsorbed energy differences are 0.58 eV, 0.05 eV, and 0.06 eV for Fe-O MOF, and 0.16 eV, 0.80 eV, and 0.75 eV for Fe-S MOF, which suggests that Fe-O possesses weak adsorption energy to OOH, Fe-S displays strong adsorption strength to OH. These adsorbed energy differences could confirm the rate-limiting steps, which are the first step for Fe-NH and Fe-O MOFs, and the fourth step for Fe-S MOF. The results confirm the Sabatier principle [41]. In other words, the Fe atom in Fe-NH MOF loses 1.26 e, in Fe-O MOF it loses 1.51 e, and in Fe-S MOF it loses 0.58 e (Table 1), thus the active Fe exhibits different adsorption energies for various ORR species, which led to the different

ORR catalysis activity. In summary, the coordinated ligand can promote the charge transfer between the central Fe atom and graphene nanosheets, and further regulate their chemical catalysis performance for HER, OER, and ORR.

3. Methods

The first-principles calculations were performed with the context of spin-polarized DFT as implemented in the Vienna ab initio simulation package (VASP) [42]. The exchange-correlation approximation was described by the generalized gradient approximation with the Perdew–Burke–Ernzerhof function [43,44]. The plane-wave cutoff energy was 500 eV for the projected augmented wave approach [45]. The vacuum layer was larger than 15 Å. All geometry structures were allowed to fully relax until the Hellmann–Feynman force on atoms was less than 0.01 eV Å^{-1}, and the total energy change was less than 1.0×10^{-5} eV. The grid density of k-point mesh in Monkhorst–Pack scheme was less than $2\pi \times 0.01$ Å^{-1}. The ab initio molecular dynamics (AIMD) were conducted with the Nosé algorithm in the NVT ensemble to investigate thermodynamical stability [46]. To treat the exchange–correlation energy of the localized d-orbital of Fe atoms, the PBE+U calculations were employed by adding the Hubbard term to the Hamiltonian [29,47]. The DFT-D3 correction was adopted to describe the long-range van der Waals interaction [48]. The VASPKIT code was used to analyze the output files from the VASP. The free-energy change (ΔG) for each fundamental step was calculated by the equation [38],

$$\Delta G = \Delta E + \Delta E_{ZPE} - T\Delta S + \Delta G_U + \Delta G_{pH} + \Delta G_{field} \tag{1}$$

where ΔE is the electronic energy difference, ΔE_{ZPE} is the zero-point energy (ZPE), T is 298.15 K, and ΔS is the difference in entropy. $\Delta G_U = eU$, where U is the electrode potential, and e is the electron transfer. $\Delta G_{pH} = k_B T \times \ln 10 \times pH$, where k_B is the Boltzmann constant, and pH = 0 in this study, which is the same as previous works [24,31,35,36]; ΔG_{field} is neglected. The entropy and vibrational frequencies of the gas species are taken from the database [49].

4. Conclusions

To summarize, we have obtained the atomic geometry, stability, HER, OER, and ORR electrocatalytic activity of Fe-MOF by using first-principles calculations. The Fe-MOFs with NH, O, and S ligands possess high stability and are atomically thin like 2D graphene. It was found that the coordinated ligands (NH, O, and S) could promote the charge redistribution in Fe-MOF, and further regulated their electronic structures and chemical performance. These three Fe-MOFs possess spin magnetism. The calculated free energy diagrams indicate that the Fe-O MOF displays the smallest hydrogen adsorption free energy ($\Delta G_H = 0.08$ eV), the Fe-NH MOF exhibits the best ORR catalysis performance with the overpotential of 0.38 V. Unfortunately, these three Fe-MOFs possess poor OER properties due to the $\eta^{OER} > 0.92$ V. Our computational results offer not only a promising strategy for the design of high efficiency versatile electrocatalysts, but also promote the following experimental exploration on the use of 2D MOF in water splitting, fuel cells, and metal-air batteries.

Supplementary Materials: The following supporting information can be downloaded online. Figure S1 supercell (2×2) of (a) Fe-NH-MOF, (b) Fe-O-MOF, and (c) Fe-S-MOF monolayers; Figure S2 the optimized top and side views of H, OOH, O, and OH on Fe-NH MOF monolayer; Figure S3 the optimized top and side views of H, OOH, O, and OH on Fe-O MOF monolayers; Figure S4 the optimized top and side views of H, OOH, O, and OH on Fe-S MOF monolayers; Table S1 optimized lattice constants (l_a) vs total energies (E_{tot}) of Fe-NH-MOF, Fe-O-MOF, and Fe-S-MOF monolayers.

Author Contributions: Conceptualization, methodology and analysis, writing—original draft preparation, review and editing, X.Y., Z.F. and Z.G.; funding acquisition, Z.F. and Z.G. All authors have read and agreed to the published version of the manuscript.

Funding: This research was funded by the Natural Science Foundation of Henan Province (No. 202300410100), key scientific research projects of Colleges and Universities in Henan Province (No. 22A140017), and Students' innovation and entrepreneurship training program of Henan Province (S202111329008 and S202111329010).

Institutional Review Board Statement: Not applicable.

Informed Consent Statement: Not applicable.

Data Availability Statement: Not applicable.

Acknowledgments: These calculations were performed in the HPCC of the national supercomputing center in Zhengzhou.

Conflicts of Interest: The authors declare no conflict of interest.

Sample Availability: Samples are available from the authors.

References

1. Seh, Z.W.; Kibsgaard, J.; Dickens, C.F.; Chorkendorff, I.; Norskov, J.K.; Jaramillo, T.F. Combining theory and experiment in electrocatalysis: Insights into materials design. *Science* **2017**, *355*, eaad4998. [CrossRef] [PubMed]
2. Chu, S.; Majumdar, A. Opportunities and challenges for a sustainable energy future. *Nature* **2012**, *488*, 294–303. [CrossRef] [PubMed]
3. Fan, J.; Chen, M.; Zhao, Z.; Zhang, Z.; Ye, S.; Xu, S.; Wang, H.; Li, H. Bridging the gap between highly active oxygen reduction reaction catalysts and effective catalyst layers for proton exchange membrane fuel cells. *Nat. Energy* **2021**, *6*, 475–486. [CrossRef]
4. Turner, J.A. Sustainable hydrogen production. *Science* **2004**, *305*, 972–974. [CrossRef] [PubMed]
5. Paul, R.; Zhu, L.; Chen, H.; Qu, J.; Dai, L. Recent advances in carbon-based metal-free electrocatalysts. *Adv. Mater.* **2019**, *31*, e1806403. [CrossRef] [PubMed]
6. Qiao, B.; Wang, A.; Yang, X.; Allard, L.F.; Jiang, Z.; Cui, Y.; Liu, J.; Li, J.; Zhang, T. Single-atom catalysis of CO oxidation using Pt1/FeOx. *Nat. Chem.* **2011**, *3*, 634–641. [CrossRef] [PubMed]
7. Wu, L.; Hu, S.; Yu, W.; Shen, S.; Li, T. Stabilizing mechanism of single-atom catalysts on a defective carbon surface. *Npj Comput. Mater.* **2020**, *6*, 23. [CrossRef]
8. Zhang, X.; Zhang, M.; Deng, Y.; Xu, M.; Artiglia, L.; Wen, W.; Gao, R.; Chen, B.; Yao, S.; Zhang, X.; et al. A stable low-temperature H2-production catalyst by crowding Pt on alpha-MoC. *Nature* **2021**, *589*, 396–401. [CrossRef]
9. Fang, S.; Zhu, X.; Liu, X.; Gu, J.; Liu, W.; Wang, D.; Zhang, W.; Lin, Y.; Lu, J.; Wei, S.; et al. Uncovering near-free platinum single-atom dynamics during electrochemical hydrogen evolution reaction. *Nat. Commun.* **2020**, *11*, 1029. [CrossRef]
10. Cao, L.; Luo, Q.; Chen, J.; Wang, L.; Lin, Y.; Wang, H.; Liu, X.; Shen, X.; Zhang, W.; Liu, W.; et al. Dynamic oxygen adsorption on single-atomic Ruthenium catalyst with high performance for acidic oxygen evolution reaction. *Nat. Commun.* **2019**, *10*, 4849. [CrossRef]
11. Zhou, Y.; Gao, G.; Kang, J.; Chu, W.; Wang, L.W. Transition metal-embedded two-dimensional C3N as a highly active electrocatalyst for oxygen evolution and reduction reactions. *J. Mater. Chem. A* **2019**, *7*, 12050–12059. [CrossRef]
12. Zhong, W.; Qiu, Y.; Shen, H.; Wang, X.; Yuan, J.; Jia, C.; Bi, S.; Jiang, J. Electronic spin moment as a catalytic descriptor for Fe single-atom catalysts supported on C_2N. *J. Am. Chem. Soc.* **2021**, *143*, 4405–4413. [CrossRef] [PubMed]
13. Huang, P.; Liu, W.; He, Z.; Xiao, C.; Yao, T.; Zou, Y.; Wang, C.; Qi, Z.; Tong, W.; Pan, B.; et al. Single atom accelerates ammonia photosynthesis. *Sci. China Chem.* **2018**, *61*, 1187–1196. [CrossRef]
14. Yang, H.; Lin, Q.; Wu, Y.; Li, G.; Hu, Q.; Chai, X.; Ren, X.; Zhang, Q.; Liu, J.; He, C. Highly efficient utilization of single atoms via constructing 3D and free-standing electrodes for CO_2 reduction with ultrahigh current density. *Nano Energy* **2020**, *70*, 104454. [CrossRef]
15. Wang, S.; Li, J.; Li, Q.; Bai, X.; Wang, J. Metal single-atom coordinated graphitic carbon nitride as an efficient catalyst for CO oxidation. *Nanoscale* **2020**, *12*, 364–371. [CrossRef] [PubMed]
16. Wang, M.; Dong, R.; Feng, X. Two-dimensional conjugated metal-organic frameworks (2D c-MOFs): Chemistry and function for MOF tronics. *Chem. Soc. Rev.* **2021**, *50*, 2764–2793. [CrossRef]
17. Yang, S.; Yu, Y.; Gao, X.; Zhang, Z.; Wang, F. Recent advances in electrocatalysis with phthalocyanines. *Chem. Soc. Rev.* **2021**, *50*, 12985–13011. [CrossRef]
18. Wei, Y.S.; Zhang, M.; Zou, R.; Xu, Q. Metal-organic framework-based catalysts with single metal sites. *Chem. Rev.* **2020**, *120*, 12089–12174. [CrossRef]
19. Wu, Z.; Adekoya, D.; Huang, X.; Kiefel, M.J.; Xie, J.; Xu, W.; Zhang, Q.; Zhu, D.; Zhang, S. Highly conductive two-dimensional metal-organic frameworks for resilient lithium storage with superb rate capability. *ACS Nano* **2020**, *14*, 12016–12026. [CrossRef]
20. Zhou, Y.; Yan, P.; Zhang, S.; Zhang, Y.; Chang, H.; Zheng, X.; Jiang, J.; Xu, Q. CO_2 coordination-driven top-down synthesis of a 2D non-layered metal–organic framework. *Fundam. Res.* **2021**, *in press*. [CrossRef]
21. Cui, Q.; Qin, G.; Wang, W.; Geethalakshmi, K.R.; Du, A.; Sun, Q. Mo-based 2D MOF as a highly efficient electrocatalyst for reduction of N_2 to NH_3: A density functional theory study. *J. Mater. Chem. A* **2019**, *7*, 14510–14518. [CrossRef]

22. Feng, Z.; Yang, Z.; Meng, X.; Li, F.; Guo, Z.; Zheng, S.; Su, G.; Ma, Y.; Tang, Y.; Dai, X. Two-dimensional metal-organic frameworks Mo3(C2O)12 as promising single-atom catalysts for selective nitrogen-to-ammonia. *J. Mater. Chem. A* **2022**. [CrossRef]

23. Zhang, K.; Guo, W.; Liang, Z.; Zou, R. Metal-organic framework based nanomaterials for electrocatalytic oxygen redox reaction. *Sci. China Chem.* **2019**, *62*, 417–429. [CrossRef]

24. Fan, X.; Tan, S.; Yang, J.; Liu, Y.; Bian, W.; Liao, F.; Lin, H.; Li, Y. From theory to experiment: Cascading of thermocatalysis and electrolysis in oxygen evolution reactions. *ACS Energy Lett.* **2021**, *7*, 343–348. [CrossRef]

25. Yu, M.; Dong, R.; Feng, X. Two-dimensional carbon-rich conjugated frameworks for electrochemical energy applications. *J. Am. Chem. Soc.* **2020**, *142*, 12903–12915. [CrossRef]

26. Feng, Z.; Li, Y.; Ma, Y.; An, Y.; Dai, X. Magnetic and electronic properties of two-dimensional metal-organic frameworks TM3(C2NH)12. *Chin. Phys. B* **2021**, *30*, 097102. [CrossRef]

27. Dong, R.; Zhang, Z.; Tranca, D.C.; Zhou, S.; Wang, M.; Adler, P.; Liao, Z.; Liu, F.; Sun, Y.; Shi, W.; et al. A coronene-based semiconducting two-dimensional metal-organic framework with ferromagnetic behavior. *Nat. Commun.* **2018**, *9*, 2637. [CrossRef]

28. Roy, P.; Pramanik, A.; Sarkar, P. Graphitic carbon nitride sheet supported single-atom metal-free photocatalyst for oxygen reduction reaction: A first-principles analysis. *J. Phys. Chem. Lett.* **2021**, *12*, 2788–2795. [CrossRef]

29. Chanier, T.; Sargolzaei, M.; Opahle, I.; Hayn, R.; Koepernik, K. LSDA+U versus LSDA: Towards a better description of the magnetic nearest-neighbor exchange coupling in Co- and Mn-doped ZnO. *Phys. Rev. B* **2006**, *73*, 134418. [CrossRef]

30. Kattel, S.; Atanassov, P.; Kiefer, B. Stability, electronic and magnetic properties of in-plane defects in graphene: A first-principles study. *J. Phys. Chem. C* **2012**, *116*, 8161–8166. [CrossRef]

31. Dang, Q.; Lin, H.; Fan, Z.; Ma, L.; Shao, Q.; Ji, Y.; Zheng, F.; Geng, S.; Yang, S.Z.; Kong, N.; et al. Iridium metallene oxide for acidic oxygen evolution catalysis. *Nat. Commun.* **2021**, *12*, 6007. [CrossRef] [PubMed]

32. Zeng, H.; Liu, X.; Chen, F.; Chen, Z.; Fan, X.; Lau, W. Single atoms on a nitrogen-doped boron phosphide monolayer: A new promising bifunctional electrocatalyst for ORR and OER. *ACS Appl. Mater. Interfaces* **2020**, *12*, 52549–52559. [CrossRef] [PubMed]

33. Li, B.; Wu, Y.; Li, N.; Chen, X.; Zeng, X.; Arramel; Zhao, X.; Jiang, J. Single-metal atoms supported on MBenes for robust electrochemical hydrogen evolution. *ACS Appl. Mater. Interfaces* **2020**, *12*, 9261–9267. [CrossRef] [PubMed]

34. Wang, C.; Liu, Y.; Yuan, J.; Wu, P.; Zhou, W. Scaling law of hydrogen evolution reaction for InSe monolayer with 3d transition metals doping and strain engineering. *J. Energy Chem.* **2020**, *41*, 107–114. [CrossRef]

35. Chen, X.; Sun, F.; Bai, F.; Xie, Z. DFT study of the two dimensional metal–organic frameworks X3(HITP)2 as the cathode electrocatalysts for fuel cell. *Appl. Surf. Sci.* **2019**, *471*, 256–262. [CrossRef]

36. Qi, S.; Wang, J.; Song, X.; Fan, Y.; Li, W.; Du, A.; Zhao, M. Synergistic trifunctional electrocatalysis of pyridinic nitrogen and single transition-metal atoms anchored on pyrazine-modified graphdiyne. *Sci. Bull.* **2020**, *65*, 995–1002. [CrossRef]

37. Zhao, C.X.; Liu, J.N.; Wang, J.; Ren, D.; Li, B.Q.; Zhang, Q. Recent advances of noble-metal-free bifunctional oxygen reduction and evolution electrocatalysts. *Chem. Soc. Rev.* **2021**, *50*, 7745–7778. [CrossRef]

38. Nørskov, J.K.; Rossmeisl, J.; Logadottir, A.; Lindqvist, L.; Kitchin, J.R.; Bligaard, T.; Jónsson, H. Origin of the overpotential for oxygen reduction at a fuel-cell cathode. *J. Phys. Chem. B* **2014**, *108*, 17886–17892. [CrossRef]

39. Feng, Z.; Li, R.; Ma, Y.; Li, Y.; Wei, D.; Tang, Y.; Dai, X. Molecule-level graphdiyne coordinated transition metals as a new class of bifunctional electrocatalysts for oxygen reduction and oxygen evolution reactions. *Phys. Chem. Chem. Phys.* **2019**, *21*, 19651–19659. [CrossRef]

40. Xu, H.; Cheng, D.; Cao, D.; Zeng, X.C. A universal principle for a rational design of single-atom electrocatalysts. *Nat. Catal.* **2018**, *1*, 339–348. [CrossRef]

41. Chorkendorff, I.; Niemantsverdriet, J.W. *Concepts of Modern Catalysis and Kinetics*; Wiley: New York, NY, USA, 2007.

42. Kresse, G.; Furthmüller, J. Efficient iterative schemes for ab initio total-energy calculations using a plane-wave basis set. *Phys. Rev. B* **1996**, *54*, 11169–11186. [CrossRef] [PubMed]

43. Perdew, J.P.; Chevary, J.A.; Vosko, S.H.; Jackson, K.A.; Pederson, M.R.; Singh, D.J.; Fiolhais, C. Atoms, molecules, solids, and surfaces: Applications of the generalized gradient approximation for exchange and correlation. *Phys. Rev. B* **1992**, *46*, 6671–6687. [CrossRef] [PubMed]

44. Perdew, J.P.; Burke, K.; Ernzerhof, M. Generalized gradient approximation made simple. *Phys. Rev. Lett.* **1996**, *77*, 3865–3868. [CrossRef]

45. Pe, B. Projector augmented-wave method. *Phys. Rev. B* **1994**, *50*, 17953–17979.

46. Martyna, G.J.; Klein, M.L.; Tuckerman, M. Nosé–hoover chains: The canonical ensemble via continuous dynamics. *J. Chem. Phys.* **1992**, *97*, 2635–2643. [CrossRef]

47. Stoeffler, D.; Etz, C. Ab initioelectronic structure and magnetism in Sr2XMoO6(X = Fe or Co) double perovskite systems: A GGA and GGA+ U comparative study. *J. Phys. Condens. Mat.* **2006**, *18*, 11291–11300. [CrossRef]

48. Grimme, S. Semiempirical gga-type density functional constructed with a longrange dispersion correction. *J. Comput. Chem.* **2006**, *27*, 1787–1799. [CrossRef]

49. Linstrom, P.J.; Mallard, W.G. The NIST chemistry WebBook: A chemical data resource on the internet. *J. Chem. Eng. Data* **2001**, *46*, 1059–1063. [CrossRef]

molecules — MDPI

Article

$Sm_{0.5}Sr_{0.5}Co_{1-x}Ni_xO_{3-\delta}$—A Novel Bifunctional Electrocatalyst for Oxygen Reduction/Evolution Reactions

Xingmei Liu, Yuwei Wang *, Liquan Fan *, Weichao Zhang, Weiyan Cao, Xianxin Han, Xijun Liu and Hongge Jia

Heilongjiang Provincial Key Laboratory of Polymeric Composite Materials, College of Materials Science and Engineering, Qiqihar University, Qiqihar 161006, China; lxm19980112@163.com (X.L.); zhang980502@163.com (W.Z.); weiyan19970606@163.com (W.C.); 03498@qqhru.edu.cn (X.H.); liuxijun2002@163.com (X.L.); jiahongge@qqhru.edu.cn (H.J.)
* Correspondence: ywwang@qqhru.edu.cn (Y.W.); 02275@qqhru.edu.cn (L.F.)

Abstract: The development of non-precious metal catalysts with excellent bifunctional activities is significant for air–metal batteries. ABO_3-type perovskite oxides can improve their catalytic activity and electronic conductivity by doping transition metal elements at B sites. Here, we develop a novel $Sm_{0.5}Sr_{0.5}Co_{1-x}Ni_xO_{3-\delta}$ (SSCN) nanofiber-structured electrocatalyst. In 0.1 M KOH electrolyte solution, $Sm_{0.5}Sr_{0.5}Co_{0.8}Ni_{0.2}O_{3-\delta}$ (SSCN82) with the optimal Co: Ni molar ratio exhibits good electrocatalytic activity for OER/ORR, affording a low onset potential of 1.39 V, a slight Tafel slope of 123.8 mV dec^{-1}, and a current density of 6.01 mA cm^{-2} at 1.8 V, and the ORR reaction process was four-electron reaction pathway. Combining the morphological characteristic of SSCN nanofibers with the synergistic effect of cobalt and nickel with a suitable molar ratio is beneficial to improving the catalytic activity of SSCN perovskite oxides. SSCN82 exhibits good bi-functional catalytic performance and electrochemical double-layer capacitance.

Keywords: $Sm_{0.5}Sr_{0.5}Co_{1-x}Ni_xO_{3-\delta}$; perovskite; cathode electrocatalyst; OER/ORR

Citation: Liu, X.; Wang, Y.; Fan, L.; Zhang, W.; Cao, W.; Han, X.; Liu, X.; Jia, H. $Sm_{0.5}Sr_{0.5}Co_{1-x}Ni_xO_{3-\delta}$—A Novel Bifunctional Electrocatalyst for Oxygen Reduction/Evolution Reactions. *Molecules* **2022**, 27, 1263. https://doi.org/10.3390/molecules27041263

Academic Editors: Jingqi Guan and Yin Wang

Received: 29 December 2021
Accepted: 10 February 2022
Published: 14 February 2022

Publisher's Note: MDPI stays neutral with regard to jurisdictional claims in published maps and institutional affiliations.

1. Introduction

Due to the excessive use of fossil fuels, environmental pollution and energy shortages have become significant challenges for human survival. Therefore, it is urgent to develop novel and efficient energy conversion devices [1,2]. Natural energy sources such as wind and solar power are abundant, but their power output is subject to climate constraints [3]. Metal–air batteries have a good application prospect among many energy storage devices due to their advantages of low price, friendly environment, and good stability [4,5]. The cathode of the metal–air battery is open, and oxygen in the air can be continuously input. There are two critical electrochemical reactions at the cathode, namely oxygen reduction reaction (ORR) and oxygen evolution reaction (OER), which can determine the performance of the entire battery. Therefore, the selection of cathode catalysts is crucial. Up to now, precious metals such as Pt, IrO_2, and RuO_2 are still the most efficiently employed catalysts, but their high price and scarce content in the earth's crust also limit their practical application [6–8]. Thus, it is imperative to develop non-noble metal bifunctional catalysts with good performance [9].

For ABO_3-type perovskite oxides, the A-site is alkaline earth metals or rare earth metals with 12-fold oxygen. The B-site is a transition metal coordinated with six-fold oxygen located at the center of the octahedron. The electrochemical properties of ABO_3 perovskite oxides are significantly improved by doping the A and/or B sites, making ABO_3 one of the best candidates for non-noble metal catalysts [10–12]. Many perovskite oxides show good electrocatalytic activity. $Sm_{0.5}Sr_{0.5}CoO_{3-\delta}$ (SSC) on Vulcan XC-72R exhibited excellent bifunctional electrocatalytic performance for metal–air batteries [13]. $La_{0.5}Sr_{0.5}NiMnRu_{0.5}O_6$ (LSNMR) showed outstanding bifunctional ORR/OER activities. The ORR onset potential (E_{onset}) of LSNMR was 0.94 V, which can be the perovskite with the best ORR activity in alkaline solution so far. The OER potential was 1.66 V at 10 mA cm^{-2} [14]. Ba$_{0.5}$Sr$_{0.5}$Co$_{0.8}$Fe$_{0.2}$O$_{3-\delta}$ (BSCF) remained highly stable for OER after 1000 cycles. For BSCF9002N2, the Tafel slopes of OER and ORR are 143 and 128 mV dec^{-1}, respectively [15]. Sm$_{0.5}$Sr$_{0.5}$CoO$_{3-\delta}$ hollow nanofibers were hybridized with N-doped graphene to obtain a remarkable ORR/OER bifunctional catalyst in alkaline media [16]. Catalyst materials with various morphology and catalytic activity can be prepared by different preparation methods. Heteroatoms doping is an effective method to enhance the electrocatalytic activity of the catalysts [17].

Our previous studies [18,19] have shown that one-dimensional nanofiber-based Sm$_{0.5}$Sr$_{0.5}$CoO$_{3-\delta}$ (SSC) exhibits excellent electrochemical performance. The doping of B-site in ABO$_3$-type SSC perovskites can efficiently promote the electrocatalytic activity of the catalyst. The replacement between the two variant Ni and Co atoms in catalyst materials can generate the ligand effect, resulting in the acceleration of the charge transfer. The synergistic coupling effect of Ni^{2+} and Co^{2+} ions can afford bifunctional synergism and promote catalytic activity [20]. To the authors' knowledge, the ORR/OER activities of Ni-doped Sm$_{0.5}$Sr$_{0.5}$CoO$_{3-\delta}$ at B-site have rarely been reported. In the present work, a novel nanofiber-structured Sm$_{0.5}$Sr$_{0.5}$Co$_{1-x}$Ni$_x$O$_{3-\delta}$ was synthesized by electrospinning, and the OER and ORR catalytic activities were explored.

2. Results and Discussion

To ascertain the optimal molar ratio of Co: Ni in SSCN catalysts, SSCN82, SSCN64, SSCN55, SSCN46, and SSCN28 were prepared by electrospinning and subsequent calcination. Figure 1 shows the typical TG–DTA curve of the SSCN precursor. The thermogravimetric process of the SSCN precursor can be divided into three stages: the weight loss in the range of 0~240 °C is due to the evaporation of water in the precursor, and the weight is reduced by 13.6% [21]. The weight loss between 244 °C and 401 °C is 39.7%, because metal nitrate will decompose within this temperature range [22]. At the same time, an apparent exothermic peak of metal nitrate can be found near 265 °C. The remaining 20.57% metal nitrate is further decomposed slowly at 406~654 °C. There was no apparent phase transformation after 654 °C, indicating that the SSCN precursor had been decomposed. To ensure the complete decomposition of the precursor, the SSCN nanofibers obtained by electrostatic spinning were calcined at 800 °C.

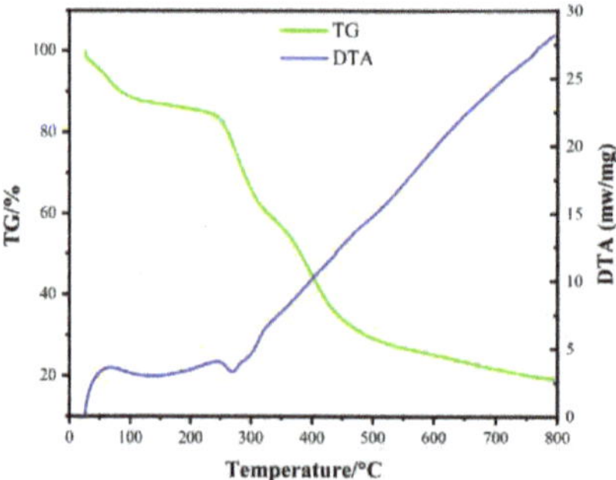

Figure 1. TG–DTA curves of the SSCN precursor with a scanning rate of 5 °C min^{-1}.

Figure 2a–e shows the SEM images of SSCN catalysts with different Co: Ni molar ratios after calcination at 800 °C for 2 h in air. It can be seen that the microstructure of SSCN fibers with different Co: Ni ratios is not the same. In comparison, the more Ni content in SSCN, the less likely it is to form a fiber structure, which may be due to the agglomeration of nickel in the calcination process. However, with the increase of Co content, SSCN82 nanofibers are not easy to fracture, and the diameter is relatively uniform. According to the literature report [16] and our previous research work [18,19], Sm$_{0.5}$Sr$_{0.5}$CoO$_{3-\delta}$ (SSC) materials prepared by electrospinning have a good nanofiber structure. While for

SSCN nanofibers, due to the intrinsic characteristics of easy agglomeration of Ni, the more nickel content in $Sm_{0.5}Sr_{0.5}Co_{1-x}Ni_xO_{3-\delta}$, the more likely the fiber will fracture or even agglomerate. The increase in Co content is beneficial to the formation of SSCN long fibers [23,24]. Thus, uniform nanofibers without obvious breakage are presented in SSCN82 with the highest Co content. Figure e1–e5 shows the EDX spectra of SSCN82 nanofibers. It clearly shows the uniform spatial distribution of Sm, Sr, Co, Ni, and O elements in the SSCN82 catalyst. In particular, the distribution of nickel can fully prove that nickel has been evenly doped into samarium strontium cobalt oxide.

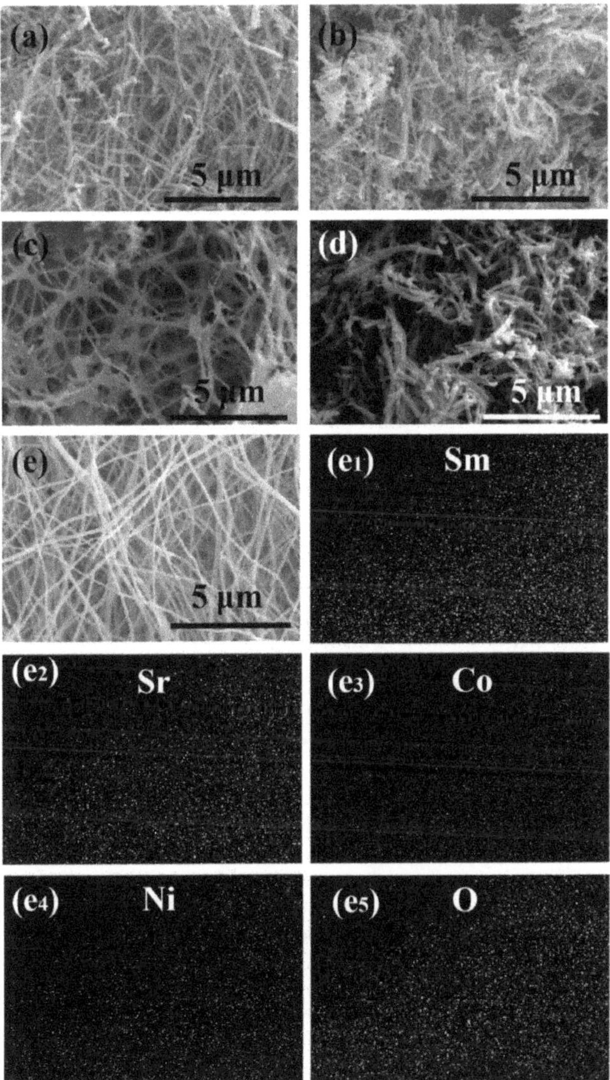

Figure 2. SEM images of (**a**) SSCN28, (**b**) SSCN46, (**c**) SSCN55, (**d**) SSCN64, and (**e**) SSCN82. (e1–e5) EDS element mappings of Figure 2e.

The wide-angle XRD patterns of the SSCN catalysts are shown in Figure 3a. The standard diffraction X-ray peaks of $SmSrCoO_3$ (PDF#53-0112) and $SmSrNiO_4$ (PDF#48-0973) are also given as guides to the eyes at the bottom of Figure 3a. All the diffraction peaks of SSCN were indexed to SmSrCoO$_3$ and SmSrNiO$_4$. The characteristic peaks appearing in Sm$_{0.5}$Sr$_{0.5}$Co$_{1-x}$Ni$_x$O$_{3-\delta}$ are consistent with the XRD pattern reported by Baek et al. [25]. The result indicates that the SSCN catalysts have a high crystalline characteristic after calcination at 800 °C. It is noteworthy that the peaks at about $2\theta \approx 24.6°$, $29.2°$, $32.2°$, and $33.4°$ are consistent with (011), (004), (013), and (110) for SmSrNiO$_4$ (PDF#48-0973), respectively. The characteristic peaks increase with the increase of Ni content for Sm$_{0.5}$Sr$_{0.5}$Co$_{1-x}$Ni$_x$O$_{3-\delta}$. Peaks at about $2\theta \approx 33.3°$, $40.9°$, and $41.2°$ can be well indexed to (121), (220), and (022) of SmSrCoO$_3$ (PDF#53-0112), respectively. With the decrease of Co content, the characteristic peak gradually weakens until SSCN28, which cannot be detected due to too little Co content. That is also why SSCN82 and SSCN64 have one peak while SSCN55, SSCN46, and SSCN28 have two peaks in the range of 30° and 35° (see Figure 3b). Moreover, from Figure 3b, the diffraction peaks of SSCN shifted towards a larger angle with the decrease of the Co/Ni ratio. Because Co ions are replaced by larger Ni ions, resulting in the lattice expansion of the SSCN unit cell [26–28].

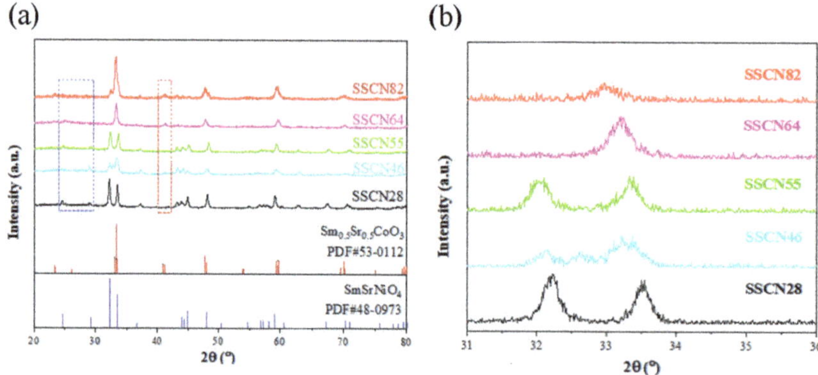

Figure 3. Wide-angle XRD patterns of SSCN82, SSCN64, SSCN55, SSCN46, and SSCN28 catalysts after calcination at 800 °C for 2 h. Standard X-ray diffraction peaks of SmSrCoO$_3$ (PDF#53-0112) and SmSrNiO$_4$ (PDF#48-0973) are given as guides to the eyes at the bottom. (**b**) A partial enlarged drawing of (**a**).

To further examine the surface composition and valence states of Co and Ni in SSCN, X-ray photoelectron spectroscopy (XPS) measurements were performed. The wide-scan spectrum of SSCN82 reveals the presence of Ni, Co, O, and Sm and Sr elements (Figure 4a). The Co 2p spectrum in Figure 4b shows the peaks of Co 2p$_{3/2}$ and Co 2p$_{1/2}$ along with their satellite peaks. For SSCN82, two peaks appear at 780.5 and 781.3 eV, which belong to Co 2p$_{3/2}$ and indicate the presence of Co^{2+} and Co^{3+}, respectively [21]. The Ni 2p spectrum (Figure 4c) shows 2p$_{3/2}$ and 2p$_{1/2}$ doublets due to spin–orbit coupling. The Ni 2p spectra of SSCN82 show two peaks at 853.8 and 859.3 eV corresponding to Ni 2p$_{3/2}$ and a satellite peak at higher binding energies [29,30]. From Figure 4d, the O 1s peaks for the sample are located at 529, 531.7, and 533.1 eV, respectively, corresponding to lattice oxygen, absorbed oxygen, and absorbed water [31,32]. The binding energies of Ni 2p and Co 2p in SSCN82 reveal that the Ni and Co atoms are uniformly distributed in the crystal structure. It further demonstrated the formation of SSCN82, which was in agreement with XRD results.

For Co oxide-based catalysts, it has been found that the number of oxidation states of Co present in the catalyst is combined with its activity for the ORR and OER. The presence of Co^{3+} is associated with higher activity towards the OER, while Co^{2+} shows higher activity towards the ORR [33]. The variability of the valence states of the cobalt ions between Co^{2+} and Co^{3+} benefits attaining excellent bifunctional ORR/OER electrocatalysis. Similar characteristics of bifunctional activity were also observed for the other transition metal oxides [34]. Ni can undergo more than one oxidation–reduction reaction during the OER [33]. The effect of Ni content on the formation of a bimetallic Co-Ni oxide electrocatalyst was explored [35]. In general, adding Ni to the catalyst can improve the half-wave ORR potential and current densities, respectively.

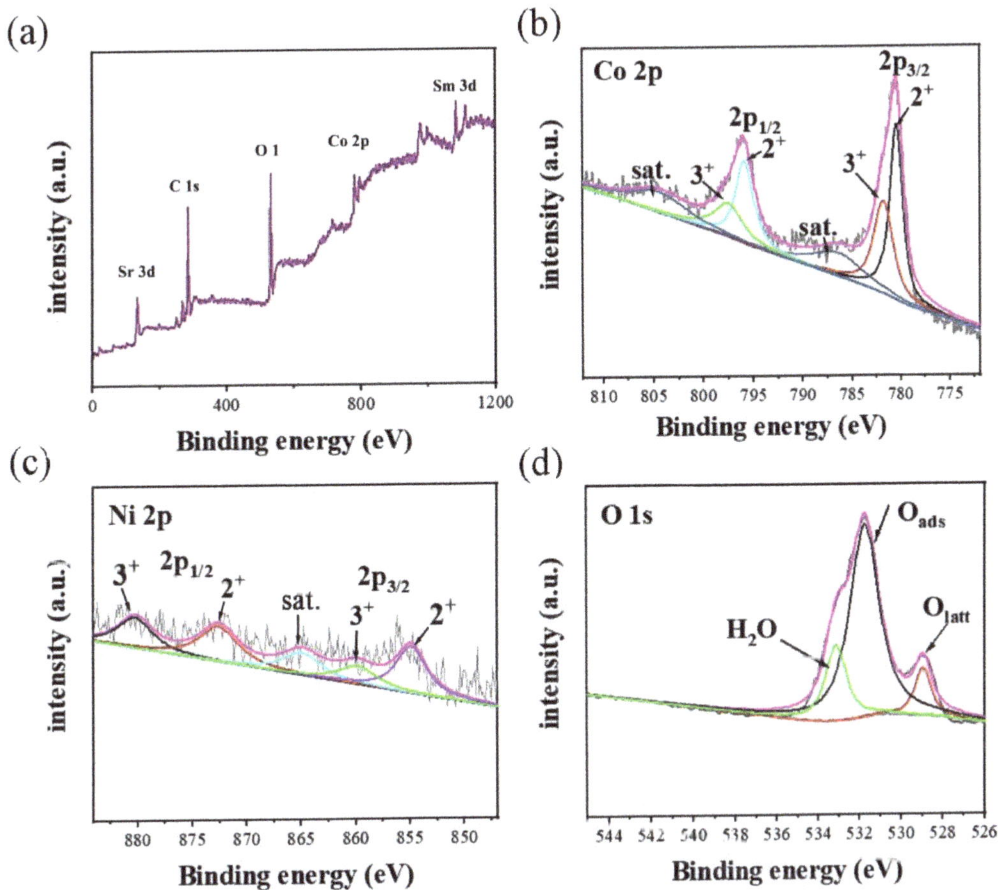

Figure 4. (**a**) XPS spectra of SSCN82 catalyst. High-resolution XPS spectra of (**b**) Co 2p, (**c**) Ni 2p, and (**d**) O 1s.

To evaluate the optimal molar ratio of cobalt and nickel of $Sm_{0.5}Sr_{0.5}Co_{1-x}Ni_xO_{3-\delta}$, the OER and ORR activities were tested at a speed of 1600 rpm in 0.1 M KOH solution. Figure 5a shows the OER polarization curves of the SSCN catalysts. SSCN82 catalyst shows a higher current density at the same potential. The onset potential (E_{onset}) of SSCN82 was 1.39 V, which was lower than the others. The results showed that SSCN82 had higher OER catalytic activity than the others. Figure 5b shows the Tafel slope obtained from the LSV curves. The OER Tafel slopes of SSCN82, SSCN64, SSCN55, SSCN46, and SSCN28 are 108.3, 137.7, 173.3, 170.4, and 180.9 mV dec^{-1}, respectively. Among the five samples, the Tafel slope of SSCN82 is the lowest, which proves that it has faster OER reaction kinetics. SSCN82 generated a current density of 6.01 mA cm^{-2} at 1.8 V. For the sake of comparison, the results in our work and the values reported in literature [15,16,36–39] are summarized in Table 1. The OER performance of SSCN82 is comparable to the perovskite catalysts.

Table 1. Comparison of OER performances in 0.1 M KOH media for SSCN catalysts with other perovskite catalysts.

Catalysts	OER Onset Potential (E/V vs. RHE)	Tafel Slope (mV dec^{-1})	Current Density at 1.8 V (mA cm^{-2})	References
SSCN82	1.39	123.8	6.01	This work
SSCN64	1.51	129.1	2.2	This work
SSCN55	1.52	133.9	1.4	This work
SSCN46	1.61	138.1	1.12	This work
SSCN28	1.63	159.2	0.65	This work
BSCF900N2		143	ca 7	[15]
SSC-HG	1.53	115		[16]
LSM	1.7	226	2	[36]
LCNP@NCNF	1.51	152	4.09	[37]
LSNF-5546	1.56	76		[38]
IrO$_2$	1.56	115	-	[39]

Furthermore, the ORR activities were also examined to demonstrate the bifunctionality of SSCN catalysts. Figure 5c shows the ORR polarization curves of SSCN at 1600 rpm. The Tafel slope (b) can be calculated by the equation: $E = a + b \log |j|$. The Tafel plots of SSCN82, SSCN64, SSCN55, SSCN46, and SSCN28 are 111.8, 95.4, 69.7 87.1, and 80.1 mV dec^{-1}, respectively. As shown in Figure 5e, the electron-transfer number of SSCN82 was closest to 4. It indicates that SSCN82 can restore O_2 to OH^- via the desired four-electron pathway [16,40]. Combining the OER and ORR test results, it is obvious that SSCN82 has the most excellent OER and ORR performance, which proved that the synergy of Co and Ni with the suitable molar ratio is beneficial for the bifunctional activities [20]. In addition, electrochemical impedance spectroscopy (EIS) analysis can further study the catalytic kinetics of SSCN. The kinetic activity of different catalysts can be expressed by their charge transfer resistance (R_{ct}). A lower R_{ct} means a faster kinetic reaction. Impedance spectra of the SSCN catalysts appeared as two capacitive arcs (Figure 5f), which can be fitted by the equivalent circuit (inset of Figure 5f). Here, R_s is the solution resistance, and CPE_1 and CPE_{dl} are two constant phase elements. R_1 represents the electron transport resistance of catalyst and electrode, and R_{ct} the interfacial charge transfer resistance. SSCN82 shows the lowest R_{ct}, indicating SSCN82 has the optimal electron and charge transport capability. The ORR/OER and CV test results divulge that a 0.8:0.2 molar proportion of Co: Ni is optimum for $Sm_{0.5}Sr_{0.5}Co_{1-x}Ni_xO_{3-\delta}$. Combining the best morphological characteristic of SSCN82 nanofibers with the synergistic effect of cobalt and nickel with the optimal molar ratio is beneficial to improving the catalytic activity of SSCN82 [20].

For oxygen electrode catalysts, the electrochemical double-layer capacitance (C_{dl}) is proportional to the active area of the catalyst [23], and the actual electrochemical active area can be calculated from CV test. Figure 6 shows the CV curves of SSCN catalysts with different cobalt–nickel ratios in the potential range of 1.20~1.30 V and at different scanning speeds (20~100 mV s^{-1}). In the potential range, the transient non-Faraday current is only caused by the structural changes of the double electric layer. Therefore, the actual surface area of the electrode can be measured by studying the adsorption and desorption behavior of the electrode surface through the non-Faraday current in this range. It can be seen from the figure that the charge and discharge current of the double layer increases linearly with the increase of scanning speed. The C_{dl} values of SSCN82, SSCN64, SSCN55, SSCN46, and SSCN28 catalysts were 1.92, 1.77, 1.71, 1.36, and 1.32 mF cm^{-2}, respectively. SSCN82 has the highest C_{dl} value and reactivity area, so it has the best catalytic activity among all SSCN catalysts.

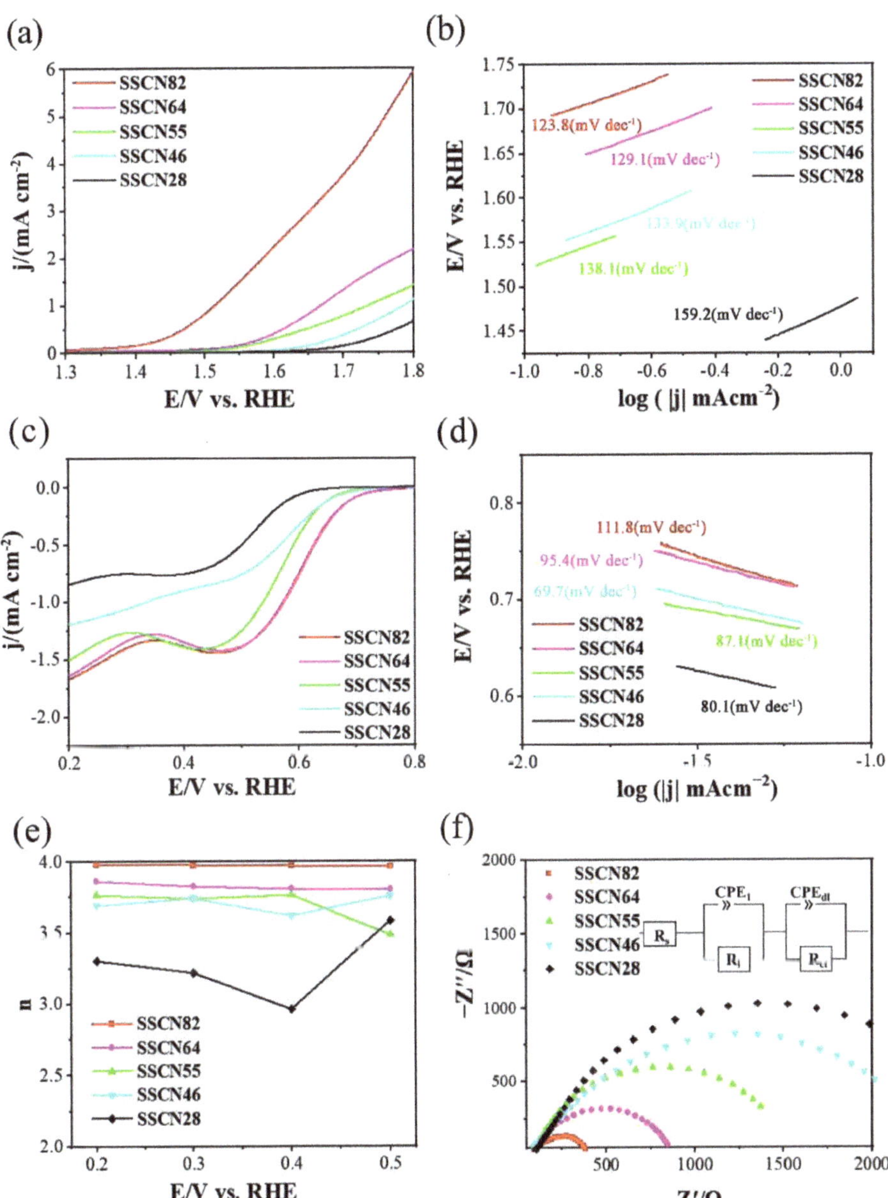

Figure 5. OER and ORR activities of SSCN. (**a**) OER polarization curves of SSCN28, SSCN64, SSCN55, SSCN46, and SSCN82 catalysts obtained in 0.1 M KOH solution with a scan rate of 5 mV s^{-1}, and (**b**) the corresponding Tafel plots of SSCN catalysts. (**c**) ORR polarization curves of SSCN28, SSCN46, SSCN55, SSCN64, and SSCN82 obtained in O$_2$-saturated 0.1 M KOH solution at 1600 rpm, and (**d**) the corresponding Tafel plots of SSCN catalysts. (**e**) Electron-transfer number (*n*) of SSCN catalysts. (**f**) Electrochemical impedance spectra at 1.664 V vs. RHE.

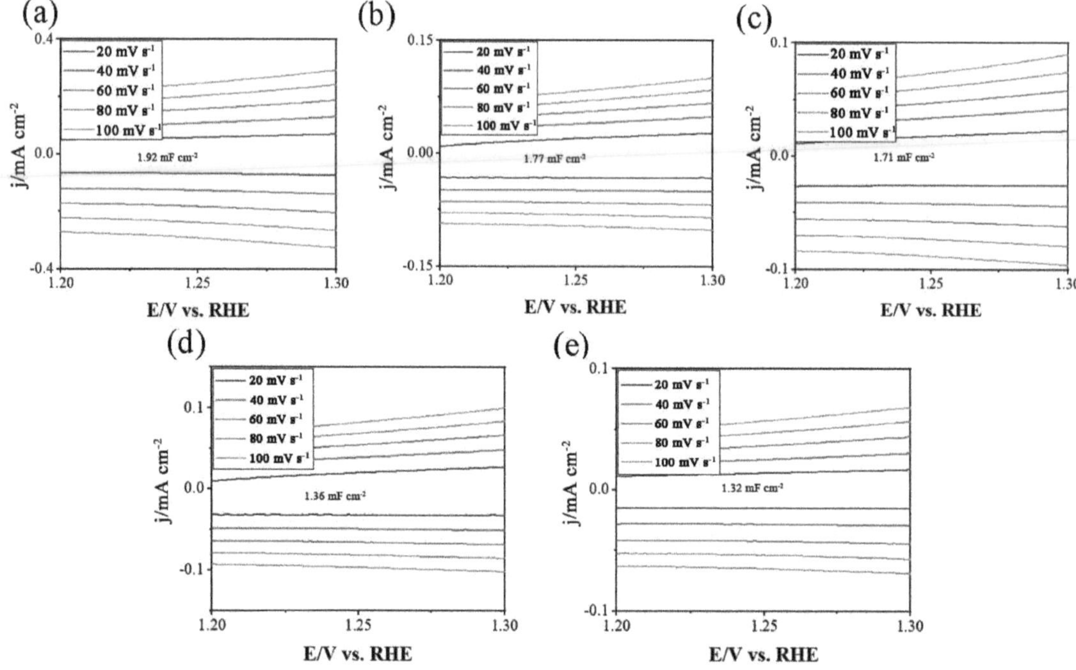

Figure 6. CV curves of (**a**) SSCN82, (**b**) SSCN64, (**c**) SSCN55, (**d**) SSCN46, and (**e**) SSCN28 catalysts obtained with different scan rates (20~100 mV s^{-1}) between a potential range of 1.20~1.30 V vs. RHE.

3. Materials and Methods

$Sm_{0.5}Sr_{0.5}Co_{1-x}Ni_xO_{3-\delta}$ (SSCN, x = 0.2, 0.4, 0.5, 0.6, and 0.8) nanofibers were synthesized by the electrospinning method. Stoichiometric amounts of samarium nitrate ($Sm(NO_3)_3 \cdot 6H_2O$), strontium nitrate ($Sr(NO_3)_2$), cobalt nitrate hexahydrate ($Co(NO_3)_2 \cdot 6H_2O$), and nickel nitrate ($Ni(NO_3)_2 \cdot 6H_2O$) with the molar ratio of 0.5:0.5:1-x:x were added into N, N-dimethyl formamide (DMF) under stirring. Then, polyvinylpyrrolidone (PVP) was dissolved into the above solution under constant stirring for several hours to form a clear and homogeneous electrospinning precursor solution. The electrospinning method was applied to synthesize the SSCN precursor nanofibers. Then the precursors were dried in a vacuum drying chamber at 150 °C for 4 h to remove the excess solvent, and subsequently calcined at 800 °C for 2 h in air with a rate of 3 °C min^{-1}, obtaining the $Sm_{0.5}Sr_{0.5}Co_{0.8}Ni_{0.2}O_{3-\delta}$, $Sm_{0.5}Sr_{0.5}Co_{0.6}Ni_{0.4}O_{3-\delta}$, $Sm_{0.5}Sr_{0.5}Co_{0.5}Ni_{0.5}O_{3-\delta}$, $Sm_{0.5}Sr_{0.5}Co_{0.4}Ni_{0.6}O_{3-\delta}$, and $Sm_{0.5}Sr_{0.5}Co_{0.2}Ni_{0.8}O_{3-\delta}$ samples. Accordingly, the samples were denoted as SSCN82, SSCN64, SSCN55, SSCN46, and SSCN28, respectively.

Thermal gravimetric analysis (TGA) was investigated by a simultaneous thermal analyzer (STA 449 F3) with a scanning rate of 5 °C min^{-1}. Scanning electron microscopy (SEM) and energy dispersion X-ray spectrometer (EDX) were operated by Hitachi S-4300 with the accelerating voltage of 10 kV. The phase composition of SSCN samples was characterized by X-ray diffraction (XRD, D 8, Germany BRUKER-AXS) with Cu-Kα radiation. The X-ray photoelectron spectroscopy (XPS) analysis was operated using an ESCALAB 250 Xi with Mg-Kα radiation.

The electrochemical measurements were manipulated by the RRDE-ALS rotate disk electrode system and CHI 760E electrochemical workstation in three-electrode mode. GCE was used as the working electrode with a diameter of 4 mm, platinum wire as the counter electrode, and Ag/AgCl as the reference electrode. A 0.1 M KOH solution was used as the electrolyte. Before electrochemical tests, oxygen was introduced into the electrolyte solution at least for 30 min to obtain the O_2-saturation solution. Linear sweep voltammetry (LSV) curves were examined at the rotating rate of 1600 rpm and the scan rate of $5\,mV\,s^{-1}$. The electron transfer number (n) during the ORR process was calculated by the following equation:

$$n = \frac{4 \times I_{disk}}{I_{disk} + I_{ring}/N}$$

where I_{disk} is the disk current, and I_{ring} is the ring current. N ($= 0.4$) is the theoretical current collection efficiency of the Pt ring. The electrochemical impedance spectroscopy (EIS) was obtained with 5 mV amplitude within 100 kHz to 0.1 Hz.

4. Conclusions

$Sm_{0.5}Sr_{0.5}Co_{1-x}Ni_xO_{3-\delta}$ (SSCN, x = 0.2, 0.4, 0.5, 0.6, and 0.8) nanofibers were successfully synthesized by electrospinning method. The synergistic effect of Co and N and the three-dimensional structure of SSCN nanofibers result in a good catalytic activity. In 0.1 M KOH electrolyte solution, SSCN82 with the optimum Co: Ni molar ratio exhibits good ORR/OER catalytic performances and electrochemical double-layer capacitance. The SSCN82 nanofibers are a promising cathode electrocatalyst for metal–air batteries.

Author Contributions: Conceptualization, Y.W. and L.F.; methodology, X.L. (Xijun Liu); formal analysis, Y.W. and L.F.; investigation, X.L. (Xingmei Liu) and W.Z.; writing—original draft preparation, X.L. (Xingmei Liu), Y.W. and L.F.; writing—review and editing, L.F.; visualization, W.C. and X.H.; supervision, X.L. (Xijun Liu) and H.J.; project administration, L.F. and Y.W.; funding acquisition, L.F. and Y.W. All authors have read and agreed to the published version of the manuscript.

Funding: This research was supported by the Natural Science Foundation of Heilongjiang Province, China (No. LH2019E092), and the Fundamental Research Funds in Heilongjiang Provincial Universities (No. 135309347).

Institutional Review Board Statement: Not applicable.

Informed Consent Statement: Not applicable.

Data Availability Statement: Data contained within the article and Supplementary Materials are available on request from the authors.

Acknowledgments: Chaojun Liu and Zhuang Wang are gratefully acknowledged for technical assistance.

Conflicts of Interest: The authors declare no conflict of interest.

Sample Availability: Samples of the compounds are not available from the authors.

References

1. Chu, S.; Majumdar, A. Opportunities and Challenges for a Sustainable Energy Future. *Nature* **2012**, *488*, 294–303. [CrossRef] [PubMed]
2. Armaroli, N.; Balzani, V. The Future of Energy Supply: Challenges and Opportunities. *Angew. Chem. Int. Ed.* **2007**, *46*, 52–66. [CrossRef] [PubMed]
3. Xu, M.; Ivey, D.G.; Xie, Z.; Qu, W. Rechargeable Zn-Air Batteries: Progress in Electrolyte Development and Cell Configuration Advancement. *J. Power Sources* **2015**, *283*, 358–371. [CrossRef]
4. Pan, J.; Xu, Y.Y.; Yang, H.; Dong, Z.; Liu, H.; Xia, B.Y. Advanced Architectures and Relatives of Air Electrodes in Zn-Air Batteries. *Adv. Sci.* **2018**, *5*, 1700691. [CrossRef] [PubMed]
5. Yu, L.; Xu, N.; Zhu, T.; Xu, Z.; Sun, M.; Geng, D. $La_{0.4}Sr_{0.6}Co_{0.7}Fe_{0.2}Nb_{0.1}O_{3-\delta}$ Perovskite Prepared by the Sol-Gel Method with Superior Performance as a Bifunctional Oxygen Electrocatalyst. *Int. J. Hydrog. Energy* **2020**, *45*, 30583–30591. [CrossRef]
6. Eskandrani, A.A.; Ali, S.M.; Al-Otaibi, H.M. Study of the Oxygen Evolution Reaction at Strontium Palladium Perovskite Electrocatalyst in Acidic Medium. *Int. J. Mol. Sci.* **2020**, *21*, 3785. [CrossRef]
7. Kim, S.; Kwon, O.; Kim, C.; Gwon, O.; Jeong, H.Y.; Kim, K.-H.; Shin, J.; Kim, G. Strategy for Enhancing Interfacial Effect of Bifunctional Electrocatalyst: Infiltration of Cobalt Nanooxide on Perovskite. *Adv. Mater. Interfaces* **2018**, *5*, 1800123. [CrossRef]
8. Cheng, J.; Zhang, M.; Jiang, Y.; Zou, L.; Gong, Y.; Chi, B.; Pu, J.; Jian, L. Perovskite $La_{0.6}Sr_{0.1}Co_{0.2}Fe_{0.8}O_3$ as an Effective Electrocatalyst for Non-Aqueous Lithium Air Batteries. *Electrochim. Acta* **2016**, *191*, 106–115. [CrossRef]
9. Jiao, Y.; Zheng, Y.; Jaroniec, M.; Qiao, S.Z. Design of Electrocatalysts for Oxygen- and Hydrogen-Involving Energy Conversion Reactions. *Chem. Soc. Rev.* **2015**, *44*, 2060–2086. [CrossRef]

10. Mohan, S.; Mao, Y. Molten Salt Synthesized Submicron Perovskite La1–XSrxCoO$_3$ Particles as Efficient Electrocatalyst for Water Electrolysis. *Front. Mater.* **2020**, *7*, 259. [CrossRef]

11. Kushwaha, H.S.; Halder, A.; Thomas, P.; Vaish, R. CaCu$_3$Ti$_4$O$_{12}$: A Bifunctional Perovskite Electrocatalyst for Oxygen Evolution and Reduction Reaction in Alkaline Medium. *Electrochim. Acta* **2017**, *252*, 532–540. [CrossRef]

12. Yu, J.; Li, C.; Li, B.; Zhu, X.; Zhang, R.; Ji, L.; Tang, D.; Asiri, A.M.; Sun, X.; Li, Q.; et al. A Perovskite La$_2$Ti$_2$O$_7$ Nanosheet as an Efficient Electrocatalyst for Artificial N$_2$ Fixation to NH$_3$ in Acidic Media. *Chem. Commun.* **2019**, *55*, 6401–6404. [CrossRef] [PubMed]

13. Velraj, S.; Zhu, J.H. Sm$_{0.5}$Sr$_{0.5}$CoO$_{3-\delta}$—A New Bi-Functional Catalyst for Rechargeable Metal-Air Battery Applications. *J. Power Sources* **2013**, *227*, 48–52. [CrossRef]

14. Retuerto, M.; Calle-Vallejo, F.; Pascual, L.; Lumbeeck, G.; Fernandez-Diaz, M.T.; Croft, M.; Gopalakrishnan, J.; Peña, M.A.; Hadermann, J.; Greenblatt, M.; et al. La$_{1.5}$Sr$_{0.5}$NiMn$_{0.5}$Ru$_{0.5}$O$_6$ Double Perovskite with Enhanced ORR/OER Bifunctional Catalytic Activity. *ACS Appl. Mater. Interfaces* **2019**, *11*, 21454–21464. [CrossRef] [PubMed]

15. Azad, U.P.; Singh, M.; Ghosh, S.; Singh, A.K.; Ganesan, V.; Singh, A.K.; Prakash, R. Facile Synthesis of BSCF Perovskite Oxide as an Efficient Bifunctional Oxygen Electrocatalyst. *Int. J. Hydrogen Energy* **2018**, *43*, 20671–20679. [CrossRef]

16. Bu, Y.; Nam, G.; Kim, S.; Choi, K.; Zhong, Q.; Lee, J.; Qin, Y.; Cho, J.; Kim, G. A Tailored Bifunctional Electrocatalyst: Boosting Oxygen Reduction/Evolution Catalysis via Electron Transfer Between N-Doped Graphene and Perovskite Oxides. *Small* **2018**, *14*, 1802767. [CrossRef]

17. Yu, M.; Yin, Z.; Yan, G.; Wang, Z.; Guo, H.; Li, G.; Liu, Y.; Li, L.; Wang, J. Synergy of Interlayer Expansion and Capacitive Contribution Promoting Sodium Ion Storage in S, N-Doped Mesoporous Carbon Nanofiber. *J. Power Sources* **2020**, *449*, 227514. [CrossRef]

18. Fan, L.; Xiong, Y.; Liu, L.; Wang, Y.; Brito, M.E. Preparation and Performance Study of One-Dimensional Nanofiber-Based Sm$_{0.5}$Sr$_{0.5}$CoO$_{3-\delta}$-Gd$_{0.2}$Ce$_{0.8}$O$_{1.9}$ Composite Cathodes for Intermediate Temperature Solid Oxide Fuel Cells. *Int. J. Electrochem. Sci.* **2013**, *8*, 11.

19. Fan, L.; Wang, Y.; Jia, Z.; Xiong, Y.; Brito, M.E. Nanofiber-Structured SSC–GDC Composite Cathodes for a LSGM Electrolyte Based IT-SOFCs. *Ceram. Int.* **2015**, *41*, 6583–6588. [CrossRef]

20. Vignesh, A.; Vajeeston, P.; Pannipara, M.; Al-Sehemi, A.G.; Xia, Y.; Kumar, G.G. Bimetallic Metal-Organic Framework Derived 3D Hierarchical NiO/Co$_3$O$_4$/C Hollow Microspheres on Biodegradable Garbage Bag for Sensitive, Selective, and Flexible Enzyme-Free Electrochemical Glucose Detection. *Chem. Eng. J.* **2022**, *430*, 133157. [CrossRef]

21. Li, L.; Song, J.; Lu, Q.; Tan, X. Synthesis of Nano-Crystalline Sm$_{0.5}$Sr$_{0.5}$Co(Fe)O$_{3-\delta}$ Perovskite Oxides by a Microwave-Assisted Sol–Gel Combustion Process. *Ceram. Int.* **2014**, *40*, 1189–1194. [CrossRef]

22. Gao, L.; Zhu, M.; Xia, T.; Li, Q.; Li, T.; Zhao, H. Ni-Doped BaFeO$_3$—Perovskite Oxide as Highly Active Cathode Electrocatalyst for Intermediate-Temperature Solid Oxide Fuel Cells. *Electrochim. Acta* **2018**, *289*, 428–436. [CrossRef]

23. Wang, Z.; Tan, S.; Xiong, Y.; Wei, J. Effect of B Sites on the Catalytic Activities for Perovskite Oxides La$_{0.6}$Sr$_{0.4}$Co$_x$Fe$_{1-X}$O$_{3-\delta}$ as Metal-Air Batteries Catalysts. *Prog. Nat. Sci. Mater. Int.* **2018**, *28*, 399–407. [CrossRef]

24. Wang, Z.; Li, M.; Liang, C.; Fan, L.; Han, J.; Xiong, Y. Effect of Morphology on the Oxygen Evolution Reaction for La$_{0.8}$Sr$_{0.2}$Co$_{0.2}$Fe$_{0.8}$O$_{3-\delta}$ Electrochemical Catalyst in Alkaline Media. *RSC Adv.* **2016**, *6*, 69251–69256. [CrossRef]

25. Baek, S.-W.; Kim, J.H.; Bae, J. Characteristics of ABO$_3$ and A$_2$BO$_4$ (ASm, Sr; BCo, Fe, Ni) Samarium Oxide System as Cathode Materials for Intermediate Temperature-Operating Solid Oxide Fuel Cell. *Solid State Ion.* **2008**, *179*, 1570–1574. [CrossRef]

26. Liu, S.; Luo, H.; Li, Y.; Liu, Q.; Luo, J.-L. Structure-Engineered Electrocatalyst Enables Highly Active and Stable Oxygen Evolution Reaction over Layered Perovskite LaSr$_3$Co$_{1.5}$Fe$_{1.5}$O$_{10-\delta}$. *Nano Energy* **2017**, *40*, 115–121. [CrossRef]

27. Da, Y.; Zeng, L.; Wang, C.; Gong, C.; Cui, L. A Simple Approach to Tailor OER Activity of Sr$_x$Co$_{0.8}$Fe$_{0.2}$O$_3$ Perovskite Catalysts. *Electrochim. Acta* **2019**, *300*, 85–92. [CrossRef]

28. Li, P.; Wei, B.; Lü, Z.; Wu, Y.; Zhang, Y.; Huang, X. La$_{1.7}$Sr$_{0.3}$Co$_{0.5}$Ni$_{0.5}$O$_{4+\delta}$ Layered Perovskite as an Efficient Bifunctional Electrocatalyst for Rechargeable Zinc-Air Batteries. *Appl. Surf. Sci.* **2019**, *464*, 494–501. [CrossRef]

29. Ashok, A.; Kumar, A.; Ponraj, J.; Mansour, S.A. Synthesis and Growth Mechanism of Bamboo like N-Doped CNT/Graphene Nanostructure Incorporated with Hybrid Metal Nanoparticles for Overall Water Splitting. *Carbon* **2020**, *170*, 452–463. [CrossRef]

30. Cabello, A.; Gayán, P.; García-Labiano, F.; de Diego, L.F.; Abad, A.; Izquierdo, M.T.; Adánez, J. Relevance of the Catalytic Activity on the Performance of a NiO/CaAl$_2$O$_4$ Oxygen Carrier in a CLC Process. *Appl. Catal. B Environ.* **2014**, *147*, 980–987. [CrossRef]

31. Liu, C.; Wang, Z.; Zong, X.; Jin, Y.; Li, D.; Xiong, Y.; Wu, G. N & S Co-Doped Carbon Nanofiber Network Embedded with Ultrafine NiCo Nanoalloy for Efficient Oxygen Electrocatalysis and Zn-Air Battery. *Nanoscale* **2020**, *12*, 9581–9589. [CrossRef] [PubMed]

32. Chen, J.; Wu, J.; Liu, Y.; Hu, X.; Geng, D. Assemblage of Perovskite LaNiO$_3$ Connected With In Situ Grown Nitrogen-Doped Carbon Nanotubes as High-Performance Electrocatalyst for Oxygen Evolution Reaction. *Phys. Status. Solidi. A* **2018**, *215*, 1800380. [CrossRef]

33. Favaro, M.; Drisdell, W.S.; Marcus, M.A.; Gregoire, J.M.; Crumlin, E.J.; Haber, J.A.; Yano, J. An Operando Investigation of (Ni–Fe–Co–Ce)O$_x$ System as Highly Efficient Electrocatalyst for Oxygen Evolution Reaction. *ACS Catal.* **2017**, *7*, 1248–1258. [CrossRef]

34. Wang, H.; Yin, F.; Li, G.; Chen, B.; Wang, Z. Preparation, Characterization and Bifunctional Catalytic Properties of MOF(Fe/Co) Catalyst for Oxygen Reduction/Evolution Reactions in Alkaline Electrolyte. *Int. J. Hydrog. Energy* **2014**, *39*, 16179–16186. [CrossRef]

35. Lambert, T.N.; Vigil, J.A.; White, S.E.; Davis, D.J.; Limmer, S.J.; Burton, P.D.; Coker, E.N.; Beechem, T.E.; Brumbach, M.T. Electrodeposited $Ni_xCo_{3-x}O_4$ Nanostructured Films as Bifunctional Oxygen Electrocatalysts. *Chem. Commun.* **2015**, *51*, 9511–9514. [CrossRef] [PubMed]
36. Wang, Q.; Xue, Y.; Sun, S.; Li, S.; Miao, H.; Liu, Z. $La_{0.8}Sr_{0.2}Co_{1-x}Mn_xO_3$ Perovskites as Efficient Bi-Functional Cathode Catalysts for Rechargeable Zinc-Air Batteries. *Electrochim. Acta* **2017**, *254*, 14–24. [CrossRef]
37. Lin, H.; Xie, J.; Zhang, Z.; Wang, S.; Chen, D. Perovskite Nanoparticles@N-Doped Carbon Nanofibers as Robust and Efficient Oxygen Electrocatalysts for Zn-Air Batteries. *J. Colloid. Interface Sci.* **2021**, *581*, 374–384. [CrossRef]
38. Wang, C.C.; Cheng, Y.; Ianni, E.; Jiang, S.P.; Lin, B. A Highly Active and Stable $La_{0.5}Sr_{0.5}Ni_{0.4}Fe_{0.6}O_{3-\delta}$ Perovskite Electrocatalyst for Oxygen Evolution Reaction in Alkaline Media. *Electrochim. Acta* **2017**, *246*, 997–1003. [CrossRef]
39. Bu, Y.; Gwon, O.; Nam, G.; Jang, H.; Kim, S.; Zhong, Q.; Cho, J.; Kim, G. A Highly Efficient and Robust Cation Ordered Perovskite Oxide as a Bifunctional Catalyst for Rechargeable Zinc-Air Batteries. *ACS Nano* **2017**, *11*, 11594–11601. [CrossRef]
40. Luo, Q.; Lin, D.; Zhan, W.; Zhang, W.; Tang, L.; Luo, J.; Gao, Z.; Jiang, P.; Wang, M.; Hao, L.; et al. Hexagonal Perovskite $Ba_{0.9}Sr_{0.1}Co_{0.8}Fe_{0.1}Ir_{0.1}O_{3-\delta}$ as an Efficient Electrocatalyst towards the Oxygen Evolution Reaction. *ACS Appl. Energy Mater.* **2020**, *3*, 7149–7158. [CrossRef]

Article

Nickel-Based Metal-Organic Frameworks as Electrocatalysts for the Oxygen Evolution Reaction (OER)

Linda Sondermann [1], Wulv Jiang [2], Meital Shviro [2,*], Alex Spieß [1], Dennis Woschko [1], Lars Rademacher [1] and Christoph Janiak [1,*]

[1] Institut für Anorganische Chemie und Strukturchemie, Heinrich-Heine-Universität Düsseldorf, 40225 Düsseldorf, Germany; linda.sondermann@hhu.de (L.S.); alex.spiess@hhu.de (A.S.); dennis.woschko@hhu.de (D.W.); lars.rademacher@hhu.de (L.R.)

[2] Forschungszentrum Jülich GmbH, Institute of Energy and Climate Research, IEK-14: Electrochemical Process Engineering, 52425 Jülich, Germany; j.wulv@fz-juelich.de

* Correspondence: m.shviro@fz-juelich.de (M.S.); janiak@uni-duesseldorf.de (C.J.); Tel.: +49-211-81-12286 (C.J.)

Abstract: The exploration of earth-abundant electrocatalysts with high performance for the oxygen evolution reaction (OER) is eminently desirable and remains a significant challenge. The composite of the metal-organic framework (MOF) $Ni_{10}Co$-BTC (BTC = 1,3,5-benzenetricarboxylate) and the highly conductive carbon material ketjenblack (KB) could be easily obtained from the MOF synthesis in the presence of KB in a one-step solvothermal reaction. The composite and the pristine MOF perform better than commercially available Ni/NiO nanoparticles under the same conditions for the OER. Activation of the nickel-cobalt clusters from the MOF can be seen under the applied anodic potential, which steadily boosts the OER performance. $Ni_{10}Co$-BTC and $Ni_{10}Co$-BTC/KB are used as sacrificial agents and undergo structural changes during electrochemical measurements, the stabilized materials show good OER performances.

Keywords: metal-organic frameworks (MOF); electrocatalysis; oxygen evolution reaction (OER); nickel; ketjenblack

Citation: Sondermann, L.; Jiang, W.; Shviro, M.; Spieß, A.; Woschko, D.; Rademacher, L.; Janiak, C. Nickel-Based Metal-Organic Frameworks as Electrocatalysts for the Oxygen Evolution Reaction (OER). *Molecules* **2022**, *27*, 1241. https://doi.org/10.3390/molecules27041241

Academic Editors: Jingyi Guan and Yin Wang

Received: 21 January 2022
Accepted: 6 February 2022
Published: 12 February 2022

Publisher's Note: MDPI stays neutral with regard to jurisdictional claims in published maps and institutional affiliations.

1. Introduction

The depletion of fossil fuels and their direct correlation in increasing global greenhouse gas emissions through their combustion show that the development of new sustainable clean energy sources is required [1–3]. A possible solution is coupling renewable energy sources like solar and wind energy with electrochemical water splitting to convert surplus electrical energy into storable hydrogen fuel [4–7]. Electrochemical water splitting consists of two half reactions, the cathodic hydrogen evolution reaction (HER; alkaline conditions: $4\,H_2O + 4\,e^- \rightarrow 2\,H_2 + 4\,OH^-$, $E° = 0.00$ V vs. RHE) and the anodic oxygen evolution reaction (OER, alkaline conditions: $4\,OH^- \rightarrow O_2 + 4\,e^- + 2\,H_2O$, $E° = 1.23$ V vs. RHE) [8,9]. The OER involves a four electron-proton coupled transfer process to generate one oxygen molecule and occurs at an applied potential (overpotential η) much higher than the theoretical equilibrium potential of $E° = 1.23$ V vs. RHE [8,10–14]. The overpotential is the difference between the applied potential and the equilibrium potential

$$\eta_{OER} = E_{RHE} - E° \,(1.23\,\text{V}) \tag{1}$$

and it is one of the key parameters on which the performance of an electrocatalyst is evaluated on [9,15,16]. Ideally a highly active electrocatalyst produces a large current density with a small overpotential [11] together with long-term stability. Another key parameter, which gives insight into the reaction mechanism and rate-determining step, is the Tafel slope (b), which can be obtained through the Tafel-equation [12,17,18].

$$\eta = a + b \times \log(j) \tag{2}$$

Up to date electrocatalysts based on noble metals, such as ruthenium and iridium, together with their oxides RuO_2 and IrO_2, are the OER-catalysts with the best performance, independently of the pH of the electrolyte [19–21]. However, the scarcity and high costs of these noble metals severely hinder a large-scale industrial application. Subsequent research has been directed towards the development of non-noble metal alternatives for the OER [22,23]. Research focuses on finding active, stable and inexpensive electrocatalysts, where especially 3d-transition-metals such as Fe, Ni and Co are of high interest [7,24,25]. Recently metal-organic frameworks (MOFs) and MOF-based electrocatalysts have been paid much attention [26,27]. MOFs are potentially porous, crystalline coordination networks with metal nodes and bridging organic ligands [28,29]. MOFs have been investigated as electrocatalysts due to their high porosity, large surface areas, diversity of composition and structure [27,30–33]. MOFs can be used directly for electrocatalytic reactions, but they have drawbacks like (i) low electrical conductivity, (ii) mass transport problems of reactants, products and electrolyte ions through their micropores, (iii) lack of stability especially in highly acidic or alkaline aqueous environments [34]. Because of those drawbacks, MOFs are often employed as precursors or sacrificial agents to construct structured carbon-based metal oxide materials as efficient electrocatalysts [27,32,35]. To increase the low electric conductivity and the electrocatalytic performance of MOFs carbon supports are added such as graphene [36], carbon nanotubes (CNTs) [21] or ketjenblack (KB) [37]. The performance of Ni-MOF electrocatalysts could be enhanced by the introduction of a second metal such as cobalt or iron [35,38–41]. Our previous work has shown that a Ni(Fe)-MOF/KB composite exhibited remarkable OER performance [34]. The MOF Ni-BTC and Ni(Fe)-BTC (BTC = 1,3,5-benzenetricarboxylate) have a fast charge transfer rate and high activity for the OER [40,42]. A mixed-metal Ni(Co)-BTC MOF has only been used as a sacrificial agent to give $NiCo_2O_4$ [35], which encouraged us to take a closer look at the activity of the pristine MOF and the composite material including KB for the OER in this work.

2. Results and Discussion

2.1. Synthesis and Characterization of the Ni-BTC Analogs

The Ni-BTC structure is identical to the Cu-BTC or HKUST-1 structure (HKUST = Hong Kong University of Science and Technology) [43] both are built by dinuclear metal(II)-secondary building units (SBUs), which are connected by BTC in a paddle-wheel fashion to a three-dimensional network of formula $[Ni_3(BTC)_2]$ [43,44]. The composite of $Ni_{10}Co$-BTC and KB, named $Ni_{10}Co$-BTC/KB was generated through a facile MOF synthesis in the presence of ketjenblack (KB) in a one-step solvothermal reaction at 170 °C for 48 h from a mixture of $Ni(NO_3)_2 \cdot 6\,H_2O$, $Co(NO_3)_2 \cdot 6\,H_2O$ (molar Ni:Co ratio 10:1), 1,3,5-benzenetricarboxylic acid (H_3BTC), 2-methylimidazole (2-MeImH), and KB in N,N-dimethylformamide (DMF) (Figure 1). $Ni_{10}Co$-BTC and $Ni_{10}Fe$-BTC were synthesized for comparison. Ni-BTC and the mixed-metal analogs are obtained as the dimethylamine adduct $[Ni_3(BTC)_2(Me_2NH)_3]$ at the axial metal position with Me_2NH being a hydrolysis product of DMF [43] (from the CHN elemental analysis data in Table S1, Supporting Information (SI)).

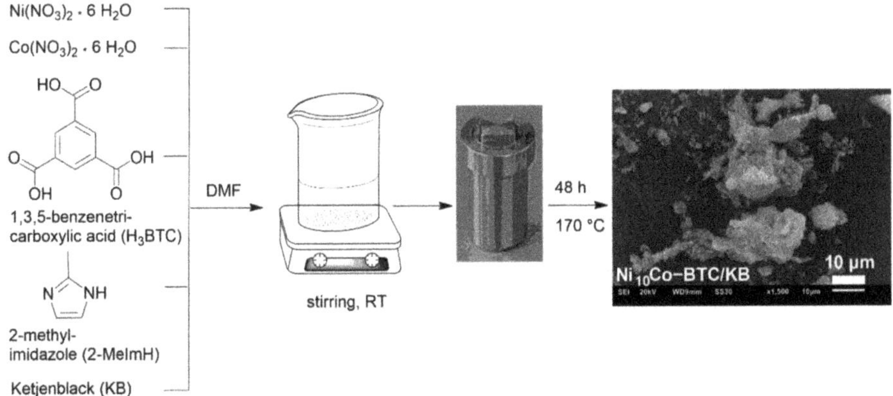

Figure 1. Schematic illustration of the Ni_{10}Co-BTC/KB composite synthesis.

A well-known phenomenon in mixed-metal MOF synthesis is that the incorporated metal ratio can differ from the starting material ratio and must be post-synthetically quantified. To quantify the amount of Co and Fe in the synthesized Ni_{10}Co-BTC and Ni_{10}Fe-BTC flame atomic absorption spectroscopy (AAS) was conducted post-synthetically, resulting in a molar Ni:Co ratio of 11:1 for Ni_{10}Co-BTC and a molar Ni:Fe ratio of 11:1 for Ni_{10}Fe-BTC (Table 1).

Table 1. SEM-EDX and AAS results of the mixed-metal samples.

Sample	SEM-EDX				AAS *	
	Molar Ratio		Metal wt.%		Approximate Molar Ratio	
	Ni	Fe/Co	Ni	Fe/Co	Ni	Fe/Co
Ni_{10}Fe-BTC	9	1	16.7	1.5	11	1
Ni_{10}Co-BTC	7	1	16.3	1.5	11	1
Ni_{10}Co-BTC/KB	7	1	10.6	1.3	8	1

* Atomic absorption spectroscopy. Weighted samples were heated to dryness with aqua regia for two times and afterwards stirred with aqua regia overnight. The solution was carefully filtered and diluted with Millipore water to a volume of 25 mL. The resulting solutions were further diluted with Millipore water (1:50) for the AAS measurements.

From the AAS determined metal wt.%, the mass fractions of the MOF in the KB composites were calculated as 67 wt.% Ni_{10}Co-BTC together with 33 wt.% KB.

Powder X-ray diffraction (PXRD) patterns of Ni_{10}Fe-BTC, Ni_{10}Co-BTC, Ni_{10}Co-BTC/KB, KB and a simulated diffraction pattern of Ni-BTC are illustrated in Figure 2a. The pristine MOF samples Ni_{10}Fe-BTC and Ni_{10}Co-BTC exhibit the same reflexes as the simulated pattern of Ni-BTC. The composite Ni_{10}Co-BTC/KB also demonstrated a PXRD pattern, which agrees to the simulated pattern of Ni-BTC and also the pristine MOF Ni_{10}Co-BTC. The aforementioned PXRD patterns show that the addition of amorphous KB in the synthesis did not influence the MOF crystal growth and structure significantly [45,46]. Apart from the reflexes of the simulated Ni-BTC MOF there are no additional reflexes in the diffraction patterns of the samples, which reveals that no iron or cobalt(oxy)hydroxides formed in the synthesis. This indicates that the second metal (iron or cobalt) was well incorporated into the structure of the Ni-BTC MOF. Pure KB displayed three broad diffraction peaks corresponding to the (100), (002), (101) planes of amorphous carbon [47,48].

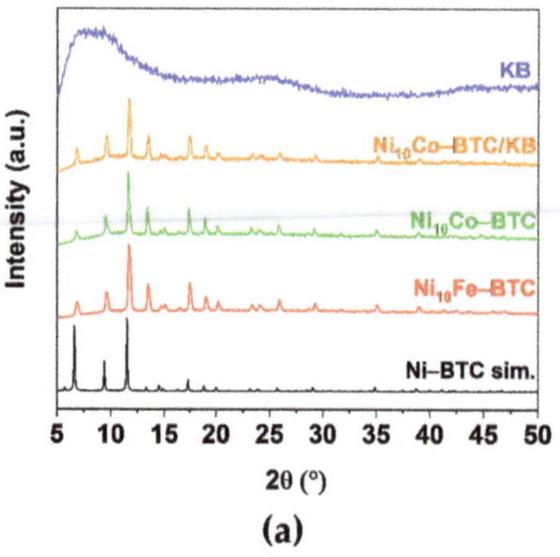
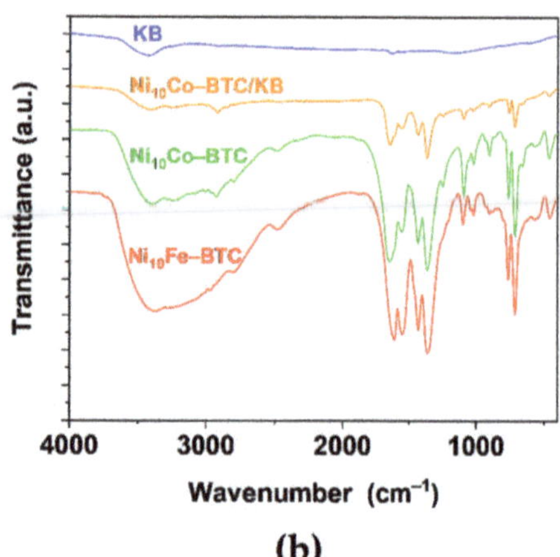

Figure 2. (**a**) PXRD patterns of experimental Ni$_{10}$Fe-BTC (red), Ni$_{10}$Co-BTC (green), Ni$_{10}$Co-BTC/KB (orange), KB (blue) and simulated Ni-BTC (black) (CCDC Nr. 802889); (**b**) FT-IR spectra of Ni$_{10}$Fe-BTC (red), Ni$_{10}$Co-BTC (green), Ni$_{10}$Co-BTC/KB (orange) and KB (blue).

Fourier transform infrared (FT-IR) spectra (Figure 2b) of the Ni-BTC analogs demonstrate the same characteristic bands (listed in Table S2, SI) which are in a good agreement with the literature [39]. A broad band in the region of 3600–3000 cm^{-1} and a band around 1650 cm^{-1} can be attributed to the stretching and bending vibrations of -OH group from adsorbed or coordinated water [43,49–51]. Characteristic vibrations of a Ni-O bond 577–460 cm^{-1} [51] and of carboxylate-groups 1617–1556 cm^{-1} (asymmetric vibration) and 1439–1364 cm^{-1} (symmetric vibration) can be observed in all samples [50,52]. Vibrations of a Fe-O bond are reported at 538 and 634 cm^{-1} and of a Fe$_2$Ni-O bond at ca. 720 cm^{-1}, which is also in the range of the vibration of a Co-O bond (725 cm^{-1}) [53–55]. Fe-O could not be detected and the Fe$_2$Ni-O and Co-O bonds are all in a similar range to each other and to aromatic vibrations (Table S2, SI).

The specific Brunauer-Emmett-Teller (BET) surface areas and pore volumes of the materials were derived from nitrogen-adsorption isotherms at 77 K (Figure 3a) and are given in Table S3, SI.

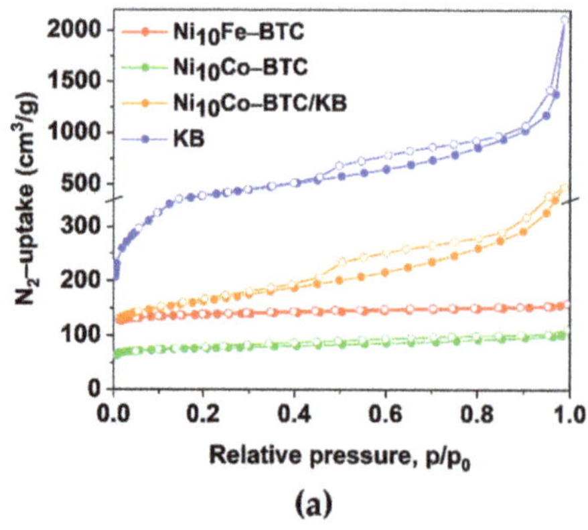
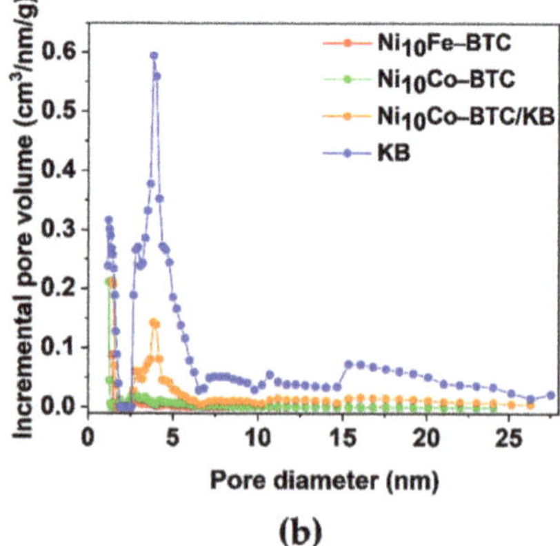

Figure 3. (**a**) N$_2$-sorption isotherms at 77 K (Adsorption: filled circles; desorption: empty circles), (**b**) pore size distributions of Ni$_{10}$Fe-BTC (red), Ni$_{10}$Co-BTC (green), Ni$_{10}$Co-BTC/KB (orange) and KB (blue).

The BET surface areas of the Ni-BTC analogs (Ni$_{10}$Fe-BTC: 555 m^2/g, Ni$_{10}$Co-BTC: 303 m^2/g and Ni$_{10}$Co-BTC/KB: 596 m^2/g) are all in the range of reported values for neat Ni-BTC (Ni-BTC$_{DMF/EtOH}$: 198 m^2/g and 252 m^2/g, Ni-BTC$_{EtOH}$: 551 m^2/g; Ni-BTC: 0.286 cm^3/g) [56,57]. Ni$_{10}$Fe-BTC and Ni$_{10}$Co-BTC show a type I isotherm revealing their microporosity with a steep gas uptake at low relative pressure followed by a plateau [58–60]. KB is a porous carbon material with a BET surface area of 1415 m^2/g, a pore volume of 1.59 cm^3/g (Table S3, SI) and mesopores, which are mostly around 4 ± 2 nm (Figure 3b). The adsorption isotherm branch of KB is a composite of a type I and II isotherm and the desorption isotherm branch additionally displays a H4 hysteresis, both being often indicators for micro-mesoporous carbons [58]. The nitrogen sorption isotherm and BET-surface area of the composite Ni$_{10}$Co-BTC/KB is a superposition of the isotherms of the MOF and KB components (Figure 3a). This superposition also holds for the pore-size distribution of the individual components in the composite (Figure 3b). The bimodal pore size distribution of Ni$_{10}$Co-BTC/KB with maxima at ~2 nm and ~4 nm reflects the contributions from the MOF and KB. Consequently, the BET surface area and pore volume of the Ni$_{10}$Co-BTC/KB composite with 596 m^2/g and 0.45 cm^3/g, respectively, are higher than the surface area and pore volume of neat Ni$_{10}$Co-BTC (303 m^2/g, 0.15 cm^3/g) but still lower than the calculated BET surface area (670 m^2/g) as determined from the sum of the mass-weighted S(BET) of KB (33 wt.%) and MOF (67 wt.%) (Equation (3) [34]):

$$S(BET)_{calc.} = \frac{\text{wt.\% of KB}}{100} \times S(BET)_{KB} + \frac{\text{wt.\% of MOF}}{100} \times S(BET)_{MOF} \tag{3}$$

The slightly lower BET surface area can be due to pore blocking effects or formation of the MOF in the mesopores of KB, as evidenced by the large reduction of the incremental mesopore volume in Figure 3b.

Thermogravimetric analyses (TGA) under N$_2$ atmosphere yield a similar curvature for all Ni-BTC analogs (Figure S1, SI). In the range of 30–320 °C the initial weight losses until decomposition can be attributed to the loss of crystal solvent molecules (DMF, H$_2$O) incorporated in the cavities [61]. After complete solvent loss the BTC-linker together with the MOF structure decomposes around 350–600 °C (Lit. 337–450 °C) (mass change of 54–57 %, Figure S1, SI) [56,60]. The TGA curves are in agreement with reported curves for NiCo-BTC and Ni-BTC [35,43,60,61].

Scanning electron microscopy (SEM) images of Ni$_{10}$Fe-BTC (Figure 4a) present spherical and cubic particles, which is in accordance with the literature [40]. Ni$_{10}$Co-BTC (Figure 4b) has irregular shaped aggregates similar to reported Ni-BTC [40]. The KB particles (Figure S3, SI) are smaller than the MOF particles, and do not have a clearly defined shape. In the composite material Ni$_{10}$Co-BTC/KB (Figure 4c) the MOF particles are covered by KB. SEM-energy-dispersive X-ray spectroscopy (EDX) was conducted for the mixed-metal MOFs (Figure 4 and Figure S2, SI) and KB (Figure S3, SI). For the mixed-metal MOFs SEM-EDX metal element mapping (Figure 4) reveals a good superposition of the two different metals. It is evident that the mapping of nickel and iron or cobalt of Ni$_{10}$Fe-BTC and Ni$_{10}$Co-BTC, respectively, is more visible than for Ni$_{10}$Co-BTC/KB, where the KB partially covers and masks surface of the MOF particles.

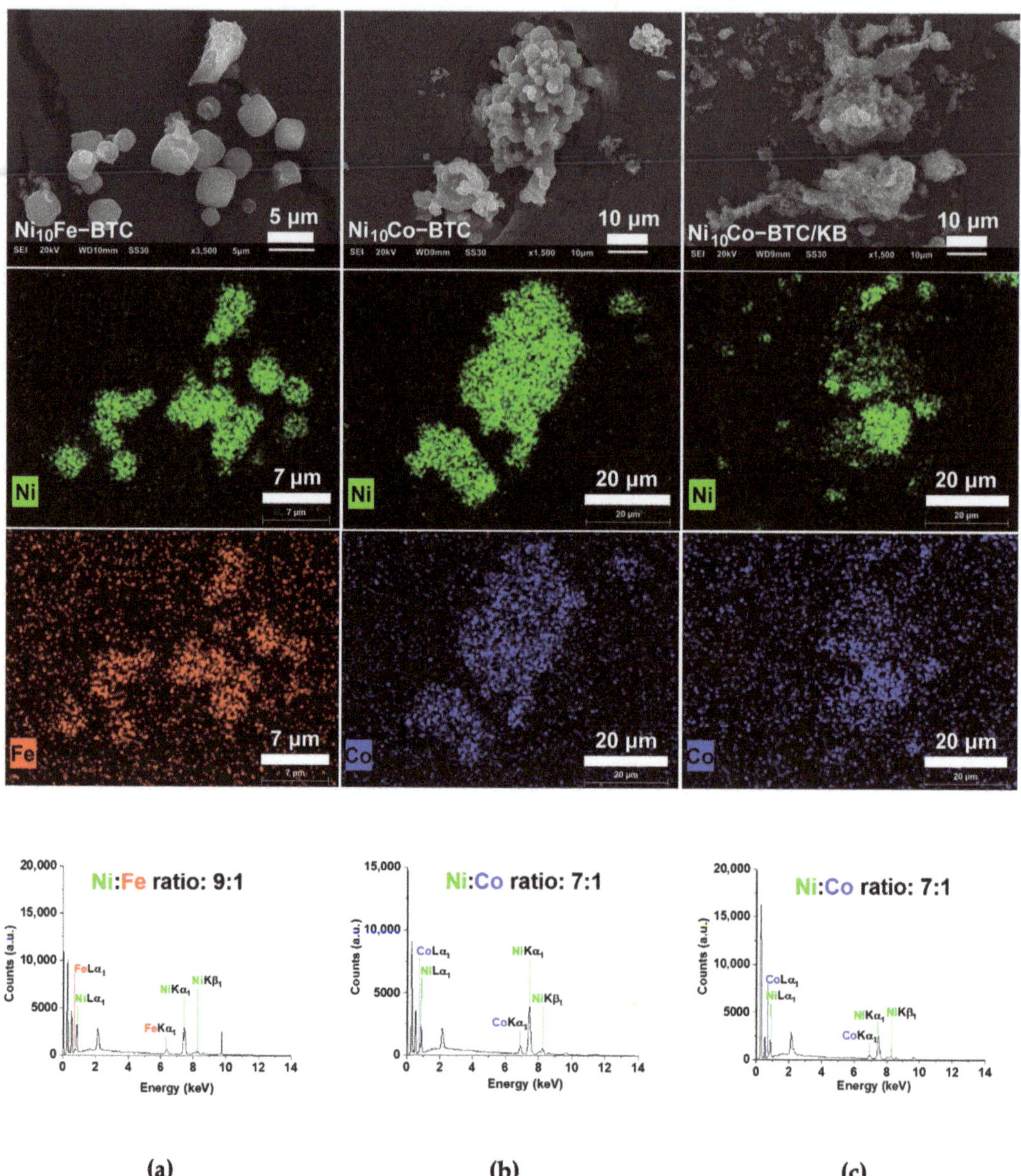

Figure 4. SEM images (**top row**), SEM-EDX metal element mappings (**two middle rows**) and EDX spectra (**bottom row**) for (**a**) $Ni_{10}Fe$-BTC, (**b**) $Ni_{10}Co$-BTC and (**c**) $Ni_{10}Co$-BTC/KB.

The metal contents of the mixed-metal samples were quantified by SEM-EDX and atomic absorption spectroscopy (AAS) and are compared in Table 1. The SEM-EDX results are more indicative of the metal ratio of the surface of the samples and the AAS results quantify the metal ratio of the bulk samples. For the synthesis of all materials a starting molar ratio of 10:1 was used for nickel to iron or cobalt. $Ni_{10}Fe$-BTC and $Ni_{10}Co$-BTC give similar AAS results with a ratio of approximately 11:1 (Ni:Fe/Co), which is close to the used 10:1 molar ratio for the synthesis. The AAS of the composite material $Ni_{10}Co$-BTC/KB results in a Ni:Co ratio of approximately 8:1, which is a little lower than the implemented molar ratio in the beginning.

2.2. Electrocatalytical Results

The OER performance of all samples was checked using a three-electrode system with rotation disk electrode in 1 mol/L KOH electrolyte. The electrochemical kinetics of the samples were evaluated by comparison of the Tafel slopes derived from linear sweep voltammetry (LSV) curves after the activation. Apart from the overpotential and Tafel slope the stability of the electrocatalysts were examined by comparing the overpotential before and after 1000 cyclic voltammetry cycles (CVs). Details are given in the experimental section.

The LSV curves in Figure 5a,b show that $Ni_{10}Fe$-BTC reaches the highest current density before the stability test in comparison to the reference commercial Ni/NiO nanoparticles, $Ni_{10}Co$-BTC, $Ni_{10}Co$-BTC/KB and KB. The LSV curves in Figure 5b display that the current density of $Ni_{10}Co$-BTC/KB was higher than that of the pristine MOF or KB before the stability test, which could be due to a conductivity enhancement effect by the introduction of KB. After the stability test the achieved current density of the pristine MOF is higher than the composite and both are higher than the current density of KB. The current densities of $Ni_{10}Fe$-BTC and of KB have declined after the stability test. $Ni_{10}Co$-BTC, $Ni_{10}Co$-BTC/KB and the commercial Ni/NiO nanoparticles reach a higher current density after the stability test. The peaks from 1.3 to 1.4 V vs. RHE in Figure 5b originated from the redox reaction of $Ni^{2+/3+}$ [35,62,63]. The redox peaks are less visible for $Ni_{10}Fe$-BTC, which is due to the well-known suppressor effect of Fe for the $Ni^{2+/3+}$ oxidation [64–66]. The changed current densities before and after the 1000 CVs already depict that an activation is taking place in case of the NiCo samples and the Ni/NiO nanoparticles. The efficiency of an electrocatalyst is normally checked with the overpotential at a current density (j) of 10 mA/cm^2, which relates to the approximate current density expected for a 10% efficient solar-to-fuel conversion device under sun illumination [13,67,68]. To have a more defined indicator the initial overpotential and the overpotential after the stability test to reach 10 mA/cm^2 should be considered. $Ni_{10}Fe$-BTC reaches 10 mA/cm^2 with an initial overpotential of 346 mV and an overpotential of 344 mV after the stability test. The measurement done after the stability test shows nearly identical values, which indicates that the material is stable in its OER performance. The results show that $Ni_{10}Fe$-BTC has a relatively good OER performance. The overpotentials needed to reach 10 mA/cm^2 before and after the stability test and Tafel slopes for KB, $Ni_{10}Co$-BTC/KB, $Ni_{10}Co$ BTC and Ni/NiO nanoparticles including the results of $Ni_{10}Fe$-BTC are listed in Table S4, SI. $Ni_{10}Co$-BTC (η = 378 mV → 337 mV), $Ni_{10}Co$-BTC/KB (η = 366 mV → 347 mV) MOF samples and the Ni/NiO nanoparticles (η = 370 mV → 358 mV) all give a decreasing overpotential, which indicates an activation of the materials and an increased OER activity. The improvement of the activity of the materials reveal that the prior activation (10 CVs) was not sufficient and it also can correlate with the formation of, for example, a highly OER active NiOOH layer [69]. Only KB demonstrated a higher overpotential afterwards (η = 376 mV → 422 mV). KB exhibits worse OER activity after 1000 cycles due to carbon corrosion at high potentials in alkaline conditions [70,71]. The carbon corrosion can also be the limiting factor of the composite, since after the 1000 CVs the $Ni_{10}Co$-BTC-derived material provides the lowest overpotential with 337 mV. The $Ni_{10}Co$-BTC/KB composite shows the best initial OER activity, but $Ni_{10}Co$-BTC has a stronger activation after the stability test and consequently a higher activity.

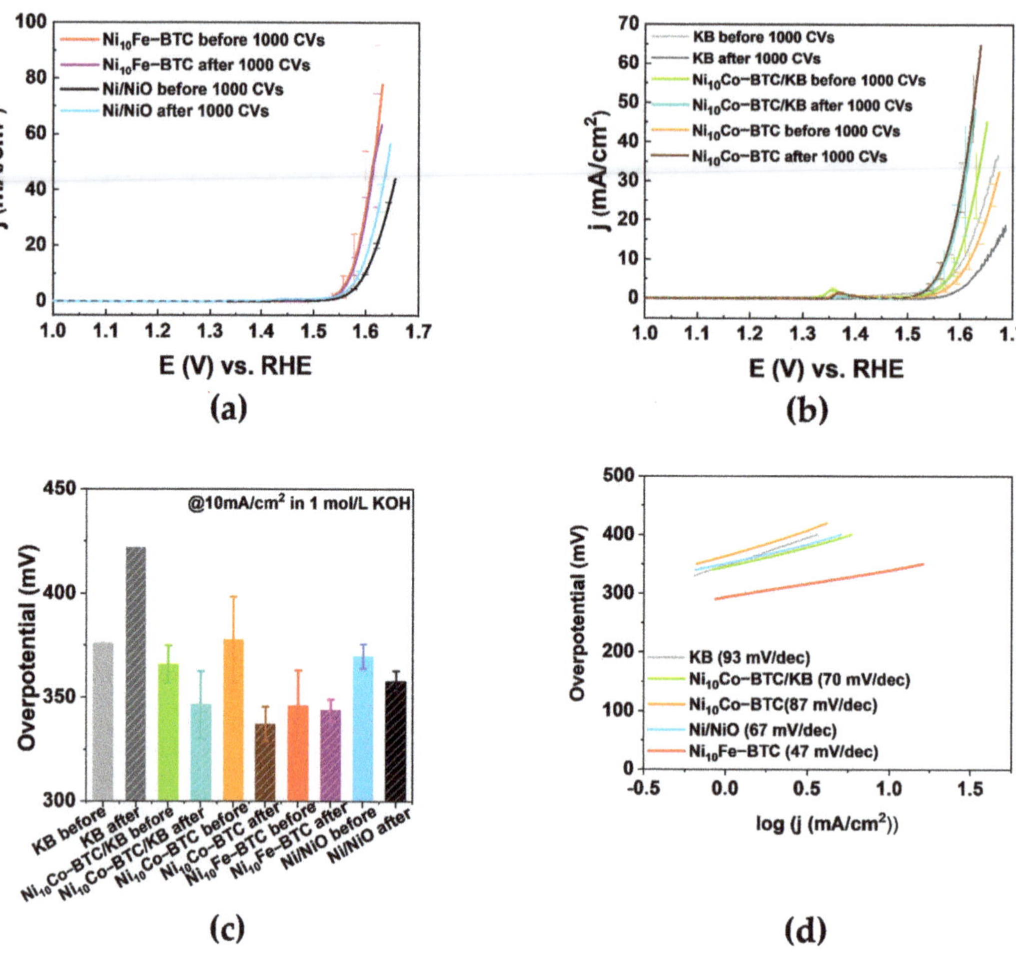

Figure 5. (**a**,**b**) LSV curves of Ni_{10}Fe-BTC, Ni/NiO nanoparticles, KB, Ni_{10}Co-BTC/KB and Ni_{10}Co-BTC before and after 1000 CVs, (**c**) overpotentials calculated from (**a**,**b**), (**d**) Tafel plots.

The Tafel slopes (Figure 5d) of Ni_{10}Fe-BTC (47 mV/dec), Ni_{10}Co-BTC/KB (70 mV/dec), Ni_{10}Co-BTC (87 mV/dec) and Ni/NiO nanoparticles (67 mV/dec) are in agreement with the reported values of Ni-BTC (71 mV/dec), $FeNi_{10}$-BTC (60 mV/dec), Fe_3Ni-BTC (49 mV/dec), FeNi-BTC (50 mV/dec) $NiCo_2O_4$ (Precursor: NiCo-BTC) (59.3 mV/dec) and $Ni(OH)_2$ (65 mV/dec) [17,35,40]. The value of the Tafel slope can give insight into the rate determining step (rds) of the OER mechanism. Krasil'shchikov's OER mechanism is one of the more widely known mechanisms, which is described by Equations (4)–(7) with the corresponding Tafel slopes b [72,73].

$$M + OH^- \rightleftarrows MOH + e^- \quad b = 120\,mV/dec \tag{4}$$

$$MOH + OH^- \rightleftarrows MO^- + e^- \quad b = 60\,mV/dec \tag{5}$$

$$MO^- \rightarrow MO + e^- \quad b = 45\,mV/dec \tag{6}$$

$$2MO \rightarrow 2M + O_2 \quad b = 19\,mV/dec \tag{7}$$

For $Ni_{10}Fe$-BTC (47 mV/dec) Equation (6) appears to be the rds of the OER mechanism. The most likely rds of the OER mechanism of $Ni_{10}Co$-BTC/KB (70 mV/dec) and Ni/NiO nanoparticles (67 mV/dec) seem to be Equation (5). The Tafel slope of $Ni_{10}Co$-BTC (87 mV/dec) is in between the values of Equations (4) and (5), which makes it difficult to clearly assign it to one of the two reaction steps.

MOFs often act as sacrificial agents to generate structured carbon-based metal oxide materials as efficient electrocatalysts [27,32,35]. To test the stability of the synthesized MOF materials in the alkaline electrolyte $Ni_{10}Co$-BTC, $Ni_{10}Co$-BTC/KB and $Ni_{10}Fe$-BTC were soaked in 1 mol/L KOH for 24 h.

The PXRD patterns of all three samples (Figure 6) display transitions of the MOF structures to their (oxy)hydroxides. $Ni_{10}Co$-BTC and $Ni_{10}Co$-BTC/KB (Figure 6a) exhibit structural changes to α-Ni(OH)$_2$ (ICDD:38-0715), β-Ni(OH)$_2$ (ICDD:14-0117), β-NiOOH (ICDD:06-0141) and/or γ-NiOOH (ICDD:06-0075) [74]. The relationship between these nickelhydroxides and oxidehydroxides is explained in the Supplementary Information. It is presently not possible, however, to quantify the components in a mixed α/β-Ni(OH)$_2$ sample from XPS results [75]. According to literature [75] α- and β-Ni(OH)$_2$ could be possibly distinguished from each other by FT-IR spectroscopy. FT-IR spectra for the samples after letting them soak in 1 mol/L KOH for 24 h only indicated also the formation of α- and β-Ni(OH)$_2$, albeit without being able to differentiate between them (Figure S7 and Table S5 in the SI). The diffraction patterns for α-Co(OH)$_2$ and γ-CoOOH match the given Ni(OH)$_2$ and NiOOH diffraction patterns [76]. Similar to $Ni_{10}Co$-BTC and $Ni_{10}Co$-BTC/KB the PXRD pattern of $Ni_{10}Fe$-BTC after 24 h in 1 mol/L KOH (Figure 6b) shows a clear loss of crystallinity of the material and indicates formation of α-Ni(OH)$_2$ (ICDD:38-0715) and/or α-FeOOH (ICDD: 29-0713) [74,77]. The change in the structure of $Ni_{10}Co$/Fe-BTC to Ni(OH)$_2$, Co(OH)$_2$ and/or to NiOOH, CoOOH and/or FeOOH is in agreement to reported observations [78]. The transition to their (oxy)hydroxides fits to the activation which could have been seen through the decrease of their overpotentials (Table S4, SI). Furthermore, transmission electron microscopy (TEM) images were made of the synthesized MOF samples before and after the electrochemical stability tests (1000 CVs). The TEM images (Figures S4–S6, SI) also indicate that a transition of the original MOF morphology takes place. The $Ni_{10}Co$-BTC/KB TEM images (Figure S6, SI) illustrate that the larger MOF particle transformed into nanoparticles. The homogenous $Ni_{10}Co$-BTC MOF particle (Figure S5, SI) changed into a carbon-based material, which contains metal (oxy)hydroxides nanoparticles. The lattice spacings of both NiCo samples (Figures S5d and S6c) could be obtained. The values of the lattice spacings fit to values of reported Ni(OH)$_2$ [79] and Co(OH)$_2$, which was formed during electrochemical tests of the MOF ZIF-67 [80]. For $Ni_{10}Fe$-BTC (Figure S4, SI) a loss of the former cubic shape of the particle can be observed and out of the resulting new morphology no lattice spacings could be gained. The changed morphology of all samples corroborates the structural changes, which could be seen through the stability test of the synthesized materials in the alkaline electrolyte.

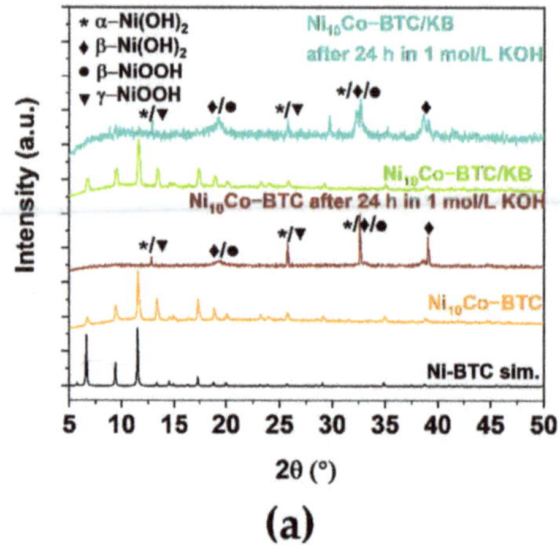

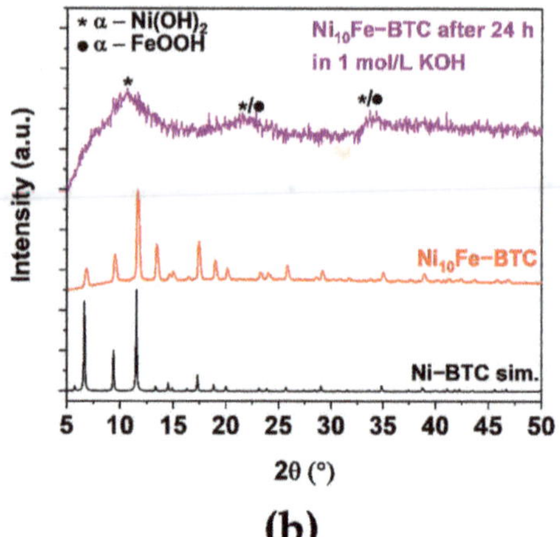

(a)　　　**(b)**

Figure 6. PXRD patterns of (**a**) experimental Ni$_{10}$Co-BTC (orange), Ni$_{10}$Co-BTC/KB (green), Ni$_{10}$Co-BTC, Ni$_{10}$Co-BTC/KB after 24 h in 1 mol/L KOH (brown and dark green) and simulated Ni-BTC (black) (CCDC Nr. 802889), (**b**) experimental Ni$_{10}$Fe-BTC (red), Ni$_{10}$Fe-BTC after 24 h in 1 mol/L KOH (purple) and simulated Ni-BTC (black) (CCDC Nr. 802889); α-Ni(OH)$_2$ marked by an asterisk (*) (ICDD: 38-0715), β-Ni(OH)$_2$ marked by a diamond (♦) (ICDD:14-0117), β-NiOOH marked by a circle (•) (ICDD:06-0141) and γ-NiOOH marked by a triangle (▼) (ICDD:06-0075) and α-FeOOH marked by a circle (•) (ICDD: 29-0713).

3. Materials and Methods

3.1. Materials

The used chemicals were obtained from commercial sources and no further purification was carried out. Ketjenblack EC 600 JD was purchased from AkzoNobel, Amsterdam. The Netherlands.

3.2. Synthesis of the Ni-BTC Analogs

Synthesis of Ni$_{10}$Fe-BTC: Ni$_{10}$Fe-BTC was synthesized according to the literature [40]. 48 mg (0.11 mmol) Fe(NO$_3$)$_3$* 9 H$_2$O, 349 mg (1.2 mmol) Ni(NO$_3$)$_2$* 6 H$_2$O, 205 mg (0.98 mmol) H$_3$BTC and 55 mg (0.67 mmol) 2-MeImH were dissolved in 15 mL DMF at room temperature (RT) and stirred for 30 min. The prepared solution was transferred into a Teflon-lined stainless-steel autoclave and then heated to 170 °C for 48 h. The resulting dark olive-green precipitate was centrifuged (25 min, 5000 rpm). The precipitate was washed one time with DMF and two times with EtOH and centrifuged again (15 min, 6000 rpm). The product was dried overnight in a vacuum drying cabinet at 90 °C and <50 mbar.

Yield (Ni$_{10}$Fe-BTC): 276 mg

Synthesis of Ni$_{10}$Co-BTC and Ni$_{10}$Co-BTC/KB: Ni$_{10}$Co-BTC and Ni$_{10}$Co-BTC/KB were synthesized according to the literature with some modifications [40]. Varying from this synthesis procedure Co(NO$_3$)$_2$* 6 H$_2$O was used instead of Fe(NO$_3$)$_3$* 9 H$_2$O for the synthesis of Ni$_{10}$Co-BTC and Ni$_{10}$Co-BTC/KB.

For the Ni$_{10}$Co-BTC sample 35 mg (0.12 mmol) Co(NO$_3$)$_2$* 6 H$_2$O, 349 mg (1.2 mmol), Ni(NO$_3$)$_2$* 6 H$_2$O, 205 mg (0.98 mmol) H$_3$BTC and 55 mg (0.67 mmol) 2-MeImH were dissolved in 20 mL DMF at RT and stirred for 30 min. For the Ni$_{10}$Co-BTC/KB sample the same amounts were used and 70 mg KB additionally added. The prepared solution was transferred into a Teflon-lined stainless-steel autoclave and then heated to 170 °C for 48 h. The resulting dark olive-green (Ni$_{10}$Co-BTC) and black (Ni$_{10}$Co-BTC/KB) precipitates were centrifuged (25 min, 5000 rpm). The precipitates were washed for one time with DMF and for two times with EtOH and centrifuged again (15 min, 6000 rpm). The products were dried overnight in a vacuum drying cabinet at 90 °C and <50 mbar.

Yield ($Ni_{10}Co$-BTC): 328 mg

Yield ($Ni_{10}Co$-BTC/KB): 357 mg

3.3. Materials Characterization

Powder X-ray diffraction (PXRD) measurements were performed at ambient temperature on a Bruker D2 Phaser powder diffractometer with a power of 300 W and an acceleration voltage of 30 kV at 10 mA using Cu-Kα radiation (λ = 1.5418 Å). The diffractograms were obtained on a low background flat silicon sample holder and evaluated with the Match 3.11 software. The samples were measured in the range from 5 to 50° 2θ with a scan speed of 2 s/step and 0.057° (2θ) step size.

Fourier transform infrared spectroscopy (FT-IR) spectra were recorded in KBr mode on a Bruker TENSOR 37 IR spectrometer in the range of 4000–400 cm^{-1}.

Nitrogen sorption measurements were performed with a Nova 4000e from Quantachrome at 77 K. The sorption isotherms were evaluated with the NovaWin 11.03 software. Prior to the measurement the materials were first degassed under vacuum (<10^{-2} mbar) at 120 °C for 5 h. Brunauer–Emmett–Teller (BET) surface areas were determined from the N_2-sorption adsorption isotherms and the pore size distributions were derived by non-local density functional theory (NLDFT) calculations based on N_2 at 77 K on carbon with slit/cylindrical pores.

Thermogravimetric analyses (TGA) were carried out with a Netzsch TG 209 F3 Tarsus device equipped with an Al crucible with a heating rate of 5 K/min under nitrogen atmosphere.

CHN elemental analyses were conducted with a Vario Mirco Cube from Elementar Analysentechnik.

Flame atomic absorption spectroscopy (AAS) was conducted with a PinAAcle 900T from PerkinElmer. Weighted samples were heated to dryness with 15 mL of aqua regia for two times and afterwards stirred with 10 mL of aqua regia overnight. The solution was carefully filtered and diluted with Millipore water to a volume of 25 mL. The resulting solutions were further diluted with Millipore water (1:50) for the AAS measurements.

Scanning electron microscopy (SEM) images were collected with a JEOL JSM-6510 LV QSEM advanced electron microscope with a LaB_6 cathode at 20 kV. The microscope was equipped with a Bruker Xflash 410 silicon drift detector and the Bruker ESPRIT software for energy-dispersive X-ray (EDX) analysis which was used to record EDX spectra and EDX mapping. The small amount of Cu, Zn and Au found in the EDX spectra are due to the brass sample holder and the sputtering of the sample with gold prior to the investigation.

Transmission electron microscopy (TEM) images of the MOF samples before the electrochemical tests were recorded on a FEI Tecnai G2 F20 electron microscope operated at 200 kV accelerating voltage equipped with a Gatan UltraScan 1000P detector. TEM samples were prepared by drop-casting the diluted material on 200 μm carbon-coated copper grids. TEM images of the samples after the electrochemical tests were obtained using a FEI Titan, 80–300 TEM with a C_s corrector for the objective lens (CEOS GmbH) operated at 300 kV. After the electrochemical test the electrode was rinsed in isopropanol and sonicated until all the layers from the surface of the electrode were dissolved into the solution. Again, the TEM samples were prepared by drop-casting the solution onto the TEM grid. Particle sizes and size distribution were determined manually using the Gatan Digital Micrograph software. For the size distribution over 150 individual particles were analyzed.

3.4. Electrocatalytic Measurements

The electrocatalytic OER measurements were conducted with a SP-150 Potentiostat form BioLogic Science Instruments and with a three-electrode setup. As reference electrode a mercury/mercury oxide (Hg/HgO) electrode was used. As counter electrode a Pt wire was used. As working electrode, a rotating disc electrode (RDE), here a glassy carbon electrode (GCE), was used. For the electrocatalyst inks 8 mg of electrocatalyst was dispersed in 1.5 mL isopropanol, 0.5 mL deionized water, and 20 μL Nafion (5 wt.%) and sonicated for 40 min. Catalyst loading was 0.2 mg/cm^2 by drop casting 10 μL ink on the GC surface (geometric area of 0.196 cm^2). All the powders dispersed well, forming a stable and homogeneous ink. After drying, the film fully covered the GC electrode. All the electrochemical measurements were conducted in 1 mol/L concentrated Ar-saturated KOH electrolyte, which has been purged with O_2 for 20 min prior to the OER experiments, with a rotation speed of 1600 rpm at RT. An activation protocol was used before the LSV measurements by cycling the working electrode between 1.0 V and 1.7 V vs. RHE at a scan rate of 100 mV/s for 10 cycles. The LSV polarization curves were recorded in a potential range of 1.0 to 1.7 V vs. RHE at a sweep rate of 5 mV/s without iR correction. The potential applied to the ohmic resistance was extracted later manually. The cycling stability was measured by comparing LSV curves before and after 1000 cycles between 1.0–1.7 V with a scan rate of 100 mV/s.

The measured potentials (vs. Hg/HgO) were converted in potentials vs. RHE with the following Equation (8) [81]:

$$E_{RHE} = E_{(Hg/HgO)} + 0.059pH + E^{\circ}{}_{(Hg/HgO)} \tag{8}$$

with E_{RHE} = converted potential vs. RHE, $E_{(Hg/HgO)}$ = measured potential and $E^{\circ}{}_{(Hg/HgO)}$ = standard potential of the Hg/HgO reference electrode.

The overpotential was calculated as shown in Equation (1): $\eta_{OER} = E_{RHE} - E^{\circ}$ (1.23 V). To reduce the experimental contingency error, at least three repeated measurements were carried out for a sample and the average curves with their error bars were compared in the figures. The OER performance of MOF samples were compared with commercial Ni/NiO nanoparticles (Alfa Aesar, Heysham, UK; VWR, International GmbH, Darmstadt, Germany) and KB (AkzoNobel, Amsterdam, The Netherlands).

4. Conclusions

Different mixed-metal Ni-BTC analogs with cobalt and iron doping were synthesized, characterized, tested for their performance in the OER and compared to the reference of Ni/NiO nanoparticles and KB. The pristine MOFs $Ni_{10}Co$-BTC and $Ni_{10}Fe$-BTC, as well as the composite $Ni_{10}Co$-BTC/KB could be prepared easily through a one-step solvothermal reaction. To compensate the shortcoming of low MOF conductivity for electrocatalysis, the highly porous and conductive carbon material KB was added, which can also support the transport of electrolyte ions and evolved gases. The MOF electrocatalysts are not stable under the implemented alkaline conditions for the electrocatalytic measurements, which again emphasizes that MOFs can be regarded as sacrificial agents. Nevertheless, the resulting, stabilized materials all evince good performances in the OER. Comparing the overpotentials of $Ni_{10}Co$-BTC (η = 378 mV) and $Ni_{10}Co$-BTC/KB (η = 366 mV) before the stability test, the composite shows a better performance for the OER, but afterward, the $Ni_{10}Co$-BTC-derived electrocatalyst exhibits a lower overpotential (337 mV) than the $Ni_{10}Co$-BTC/KB-derived electrocatalyst (347 mV). This illustrates that the conductivity, which could have been increased by introducing KB, is not the key factor limiting the OER activity of the $Ni_{10}Co$-BTC-derived electrocatalyst. However, a clearly positive effect of KB in the $Ni_{10}Co$-BTC/KB-derived material is a decreased Tafel slope with 70 mV dec^{-1} in comparison to the $Ni_{10}Co$-BTC-derived material with 87 mV dec^{-1}, which indicates a more favorable kinetics of the OER for the composite-derived material. The $Ni_{10}Fe$-BTC-dervied electrocatalyst remains the most stable material in the electrochemical OER performance (η = 346 mV $\rightarrow$ 344 mV) and has the lowest Tafel slope of 47 mV dec^{-1}, showing that the activity of Ni-electrocatalysts can be improved to some extent with incorporated iron. The results of the Tafel analysis show that the introduction of KB in the $Ni_{10}Co$-BTC MOF facilitates to overcome the kinetic barrier of the complex four electron-proton coupled OER transfer process. The composite material $Ni_{10}Co$-BTC/KB and the presented protocol give insight into the possibility of combining MOFs, as sacrificial agents, with KB to generate new MOF-based electrocatalysts for electrocatalytic reactions. Further research should now be conducted to investigate potential other nickel-metal combinations to optimize the electrocatalytic performance.

Supplementary Materials: The following supporting information can be downloaded online. Table S1. Elemental analysis of the MOF samples. Table S2. Assignments of IR-bands of Ni-BTC analogs (cm^{-1}). Table S3. BET-surface areas and total pore volumes of the Ni-BTC analogs. Table S4. Overpotentials (at 10 mA/cm^2) and Tafel slopes for Ni_{10}Fe-BTC, Ni_{10}Co-BTC, Ni_{10}Co-BTC/KB, KB and Ni/NiO nanoparticles done with a GCE (loading: 0.2 mg/cm^2) with a scan rate of 5 mV/s in a 1.0 mol/L KOH electrolyte. Figure S1. TGA curves of (a) Ni_{10}Fe-BTC, (b) Ni_{10}Co-BTC, (c) Ni_{10}Co-BTC/KB and (d) KB under N_2 atmosphere with a heating rate of 5 K/min. Figure S2. SEM-EDX spectra of (a) Ni_{10}Fe-BTC, (b) Ni_{10}Co-BTC and (c) Ni_{10}Co-BTC/KB. Figure S3. (a,b) SEM images and (c) SEM-EDX spectra of KB. Figure S4. TEM images of Ni_{10}Fe-BTC (a) before (shown particle size: 3.1 μm) and (b–d) after 1000 CVs. Figure S5. TEM images of Ni_{10}Co-BTC (a) before and (b–d) after 1000 CVs (shown particle size in (b): 2.8 μm; displayed particles in (c) give the average diameter of 20 nm ± 9 nm; (d) the lattice spacings and grain boundaries are illustrated in red and red lines, respectively). Figure S6. TEM images of Ni_{10}Co-BTC/KB (a) before (shown particle size: 4.4 μm) and (b,c) after 1000 CVs ((c) the lattice spacings and grain boundaries are illustrated in red and red lines, respectively.) and (d) histogram of Ni_{10}Co-BTC/KB after 1000 CVs determined from (b) give the average diameter of 5 nm ± 1 nm (1σ). Scheme S1. Schematic relation between β -Ni(OH)$_2$, α-Ni(OH)$_2$, β-NiOOH and γ-NiOOH. Figure S7. FT-IR spectra of (a) Ni_{10}Co-BTC after 24 h in 1 mol/L KOH (dark green) and (b) comparison with Ni_{10}Co-BTC (green), (c) Ni_{10}Co-BTC/KB after 24 h in 1 mol/L KOH (brown) and (d) comparison with Ni_{10}Co-BTC/KB (orange), (e) Ni_{10}Fe-BTC after 24 h in 1 mol/L KOH (purple) and (f) comparison with Ni_{10}Fe-BTC (red). Table S5. Assignments of IR-bands of Ni-BTC analogs after 24 h in 1 mol/L KOH (cm^{-1}). References [82–84] are cited in the supplementary materials.

Author Contributions: Conceptualization, C.J. and L.S.; Methodology, L.S.; Validation, L.S. and M.S.; Formal analysis, L.S.; Investigation, L.S., W.J., M.S., A.S., D.W. and L.R.; Resources, C.J.; Data Curation, L.S.; Writing—Original Draft, L.S.; Writing—Review & Editing, C.J., W.J. and M.S.; Visualization, L.S.; Supervision, C.J.; Project administration, C.J.; Funding acquisition, C.J. All authors have read and agreed to the published version of the manuscript.

Funding: The research of C.J. was supported by a joint National Natural Science Foundation of China–Deutsche Forschungsgemeinschaft (NSFC-DFG) project (DFG JA466/39-1).

Institutional Review Board Statement: Not applicable.

Informed Consent Statement: Not applicable.

Data Availability Statement: The data presented in this study are available on request from the corresponding author.

Conflicts of Interest: The authors declare that they have no known competing financial interest or personal relationships that could have appeared to influence the work reported in this paper.

References

1. IEA. *CO2 Emissions from Fuel Combustion 2019—Highlights*; IEA: Paris, France, 2019.
2. Özturk, S.; Xiao, Y.-X.; Dietrich, D.; Giesen, B.; Barthel, J.; Ying, J.; Yang, X.-Y.; Janiak, C. Nickel nanoparticles supported on a covalent triazine framework as electrocatalyst for oxygen evolution reaction and oxygen reduction reactions. *Beilstein J. Nanotechnol.* **2020**, *11*, 770–781. [CrossRef]
3. Seh, Z.W.; Kibsgaard, J.; Dickens, C.F.; Chorkendorff, I.; Nørskov, J.K.; Jaramillo, T.F. Combining theory and experiment in electrocatalysis: Insights into materials design. *Science* **2017**, *355*, eaad4998. [CrossRef] [PubMed]
4. Roger, I.; Shipman, M.A.; Symes, M.D. Earth-abundant catalysts for electrochemical and photoelectrochemical water splitting. *Nat. Rev. Chem.* **2017**, *1*, 3. [CrossRef]
5. Zheng, S.; Li, X.; Yan, B.; Hu, Q.; Xu, Y.; Xiao, X.; Xue, H.; Pang, H. Transition-Metal (Fe, Co, Ni) Based Metal-Organic Frameworks for Electrochemical Energy Storage. *Adv. Energy Mater.* **2017**, *7*, 1602733. [CrossRef]
6. Zhuang, L.; Ge, L.; Yang, Y.; Li, M.; Jia, Y.; Yao, X.; Zhu, Z. Ultrathin Iron-Cobalt Oxide Nanosheets with Abundant Oxygen Vacancies for the Oxygen Evolution Reaction. *Adv. Mater.* **2017**, *29*, 1606793. [CrossRef] [PubMed]

7. Han, L.; Dong, S.; Wang, E. Transition-Metal (Co, Ni, and Fe)-Based Electrocatalysts for the Water Oxidation Reaction. *Adv. Mater.* **2016**, *28*, 9266–9291. [CrossRef] [PubMed]
8. Jiao, Y.; Zheng, Y.; Jaroniec, M.; Qiao, S.Z. Design of electrocatalysts for oxygen- and hydrogen-involving energy conversion reactions. *Chem. Soc. Rev.* **2015**, *44*, 2060–2086. [CrossRef] [PubMed]
9. Tahir, M.; Pan, L.; Idrees, F.; Zhang, X.; Wang, L.; Zou, J.-J.; Wang, Z.L. Electrocatalytic oxygen evolution reaction for energy conversion and storage: A comprehensive review. *Nano Energy* **2017**, *37*, 136–157. [CrossRef]
10. Bandal, H.; Koteshwara Reddy, K.; Chaugule, A.; Kim, H. Iron-based heterogeneous catalysts for oxygen evolution reaction; change in perspective from activity promoter to active catalyst. *J. Power Sources* **2018**, *395*, 106–127. [CrossRef]
11. Wang, J.; Cui, W.; Liu, Q.; Xing, Z.; Asiri, A.M.; Sun, X. Recent Progress in Cobalt-Based Heterogeneous Catalysts for Electrochemical Water Splitting. *Adv. Mater.* **2016**, *28*, 215–230. [CrossRef]
12. Suen, N.-T.; Hung, S.-F.; Quan, Q.; Zhang, N.; Xu, Y.-J.; Chen, H.M. Electrocatalysis for the oxygen evolution reaction: Recent development and future perspectives. *Chem. Soc. Rev.* **2017**, *46*, 337–365. [CrossRef] [PubMed]
13. Wei, C.; Rao, R.R.; Peng, J.; Huang, B.; Stephens, I.E.L.; Risch, M.; Xu, Z.J.; Shao-Horn, Y. Recommended Practices and Benchmark Activity for Hydrogen and Oxygen Electrocatalysis in Water Splitting and Fuel Cells. *Adv. Mater.* **2019**, *31*, 1806296. [CrossRef] [PubMed]
14. Farid, S.; Ren, S.; Hao, C. MOF-derived metal/carbon materials as oxygen evolution reaction catalysts. *Inorg. Chem. Commun.* **2018**, *94*, 57–74. [CrossRef]
15. Anantharaj, S.; Ede, S.R.; Karthick, K.; Sankar, S.S.; Sangeetha, K.; Karthik, P.E.; Kundu, S. Precision and correctness in the evaluation of electrocatalytic water splitting: Revisiting activity parameters with a critical assessment. *Energy Environ. Sci.* **2018**, *11*, 744–771. [CrossRef]
16. Li, Y.; Du, X.; Huang, J.; Wu, C.; Sun, Y.; Zou, G.; Yang, C.; Xiong, J. Recent Progress on Surface Reconstruction of Earth- Abundant Electrocatalysts for Water Oxidation. *Small* **2019**, *15*, 1901980. [CrossRef]
17. Shinagawa, T.; Garcia-Esparza, A.T.; Takanabe, K. Insight on Tafel slopes from a microkinetic analysis of aqueous electrocatalysis for energy conversion. *Sci. Rep.* **2015**, *5*, 13801. [CrossRef]
18. Hamann, C.H.; Vielstich, W. *Elektrochemie*, 4 ed.; Wiley-VCH: Weinheim, Germany, 2005.
19. Lee, Y.; Suntivich, J.; May, K.J.; Perry, E.E.; Shao-Horn, Y. Synthesis and Activities of Rutile IrO_2 and RuO_2 Nanoparticles for Oxygen Evolution in Acid and Alkaline Solutions. *J. Phys. Chem. Lett.* **2012**, *3*, 399–404. [CrossRef] [PubMed]
20. Pi, Y.; Zhang, N.; Guo, S.; Guo, J.; Huang, X. Ultrathin Laminar Ir Superstructure as Highly Efficient Oxygen Evolution Electrocatalyst in Broad pH Range. *Nano Lett.* **2016**, *16*, 4424–4430. [CrossRef]
21. Zhao, P.; Hua, X.; Xu, W.; Luo, W.; Chen, S.; Cheng, G. Metal–organic framework-derived hybrid of Fe_3C nanorod-encapsulated, N-doped CNTs on porous carbon sheets for highly efficient oxygen reduction and water oxidation. *Catal. Sci. Technol.* **2016**, *6*, 6365–6371. [CrossRef]
22. Yang, D.; Chen, Y.; Su, Z.; Zhang, X.; Zhang, W.; Srinivas, K. Organic carboxylate-based MOFs and derivatives for electrocatalytic water oxidation. *Coord. Chem. Rev.* **2021**, *428*, 213619. [CrossRef]
23. Zhu, Y.P.; Guo, C.; Zheng, Y.; Qiao, S.-Z. Surface and Interface Engineering of Noble-Metal-Free Electrocatalysts for Efficient Energy Conversion Processes. *Acc. Chem. Res.* **2017**, *50*, 915–923. [CrossRef] [PubMed]
24. Xiao, Q.; Zhang, Y.; Guo, X.; Jing, L.; Yang, Z.; Xue, Y.; Yan, Y.-M.; Sun, K. A high-performance electrocatalyst for oxygen evolution reactions based on electrochemical post-treatment of ultrathin carbon layer coated cobalt nanoparticles. *Chem. Commun.* **2014**, *50*, 13019–13022. [CrossRef] [PubMed]
25. Saha, S.; Ganguli, A.K. FeCoNi Alloy as Noble Metal-Free Electrocatalyst for Oxygen Evolution Reaction (OER). *ChemistrySelect* **2017**, *2*, 1630–1636. [CrossRef]
26. Jin, S. How to Effectively Utilize MOFs for Electrocatalysis. *ACS Energy Lett.* **2019**, *4*, 1443–1445. [CrossRef]
27. Shao, Q.; Yang, J.; Huang, X. The Design of Water Oxidation Electrocatalysts from Nanoscale Metal–Organic Frameworks. *Chem. Eur. J.* **2018**, *24*, 15143–15155. [CrossRef]
28. Janiak, C.; Vieth, J.K. MOFs, MILs and more: Concepts, properties and applications for porous coordination networks (PCNs). *New J. Chem.* **2010**, *34*, 2366–2688. [CrossRef]
29. Batten, S.R.; Champness, N.R.; Chen, X.-M.; Garcia-Martinez, J.; Kitagawa, S.; Öhrström, L.; O'Keeffe, M.; Paik Suh, M.; Reedijk, J. Terminology of metal–organic frameworks and coordination polymers (IUPAC Recommendations 2013). *Pure Appl. Chem.* **2013**, *85*, 1715–1724. [CrossRef]
30. Aiyappa, H.B.; Masa, J.; Andronescu, C.; Muhler, M.; Fischer, R.A.; Schuhmann, W. MOFs for Electrocatalysis: From Serendipity to Design Strategies. *Small Methods* **2019**, *3*, 1800415. [CrossRef]
31. Jiao, L.; Wang, Y.; Jiang, H.-L.; Xu, Q. Metal–Organic Frameworks as Platforms for Catalytic Applications. *Adv. Mater.* **2018**, *30*, 1703663. [CrossRef]
32. Wang, H.-F.; Chen, L.; Pang, H.; Kaskel, S.; Xu, Q. MOF-derived electrocatalysts for oxygen reduction, oxygen evolution and hydrogen evolution reactions. *Chem. Soc. Rev.* **2020**, *49*, 1414–1448. [CrossRef]
33. Frydendal, R.; Paoli, E.A.; Knudsen, B.P.; Wickman, B.; Malacrida, P.; Stephens, I.E.L.; Chorkendorff, I. Benchmarking the stability of oxygen evolution reaction catalysts: The importance of monitoring mass losses. *ChemElectroChem* **2014**, *1*, 2075–2081. [CrossRef]

34. Özturk, S.; Moon, G.-H.; Spiess, A.; Budiyanto, E.; Roitsch, S.; Tüysüz, H.; Janiak, C. A Highly-Efficient Oxygen Evolution Electrocatalyst Derived from a Metal-Organic Framework and Ketjenblack Carbon Material. *ChemPlusChem* **2021**, *86*, 1106–1115. [CrossRef] [PubMed]

35. Li, Z.; Gu, A.; He, X.; Lv, H.; Wang, L.; Lou, Z.; Zhou, Q. Rod bundle-like nickel cobaltate derived from bimetal-organic coordination complex as robust electrocatalyst for oxygen evolution reaction. *Solid State Ion.* **2019**, *331*, 37–42. [CrossRef]

36. Jahan, M.; Liu, Z.; Loh, K.P. A Graphene Oxide and Copper-Centered Metal Organic Framework Composite as a Tri-Functional Catalyst for HER, OER, and ORR. *Adv. Funct. Mater.* **2013**, *23*, 5363–5372. [CrossRef]

37. Li, J.; Zhou, N.; Song, J.; Fu, L.; Yan, J.; Tang, Y.; Wang, H. Cu–MOF-Derived Cu/Cu_2O Nanoparticles and CuN_xC_y Species to Boost Oxygen Reduction Activity of Ketjenblack Carbon in Al–Air Battery. *ACS Sustain. Chem. Eng.* **2018**, *6*, 413–421. [CrossRef]

38. Deng, X.; Özturk, S.; Weidenthaler, C.; Tuysuz, H. Iron-Induced Activation of Ordered Mesoporous Nickel Cobalt Oxide Electrocatalyst for the Oxygen Evolution Reaction. *ACS Appl. Mater. Interfaces* **2017**, *9*, 21225–21233. [CrossRef]

39. Wang, L.; Wu, Y.; Cao, R.; Ren, L.; Chen, M.; Feng, X.; Zhou, J.; Wang, B. Fe/Ni Metal–Organic Frameworks and Their Binder-Free Thin Films for Efficient Oxygen Evolution with Low Overpotential. *ACS Appl. Mater. Interfaces* **2016**, *8*, 16736–16743. [CrossRef]

40. Zheng, F.; Zhang, Z.; Xiang, D.; Li, P.; Du, C.; Zhuang, Z.; Li, X.; Chen, W. Fe/Ni bimetal organic framework as efficient oxygen evolution catalyst with low overpotential. *J. Colloid Interface Sci.* **2019**, *555*, 541–547. [CrossRef]

41. Li, D.; Liu, H.; Feng, L. A Review on Advanced FeNi-Based Catalysts for Water Splitting Reaction. *Energy Fuels* **2020**, *34*, 13491–13522. [CrossRef]

42. Zhai, Z.-M.; Yang, X.-G.; Yang, Z.-T.; Lu, X.-M.; Ma, L.-F. Trinuclear Ni(II) oriented highly dense packing and π-conjugation degree of metal–organic frameworks for efficient water oxidation. *CrystEngComm* **2019**, *21*, 5862–5866. [CrossRef]

43. Maniam, P.; Stock, N. Investigation of Porous Ni-Based Metal-Organic Frameworks Containing Paddle-Wheel Type Inorganic Building Units via High-Throughput Methods. *Inorg. Chem.* **2011**, *50*, 5085–5097. [CrossRef] [PubMed]

44. Furukawa, H.; Go, Y.B.; Ko, N.; Park, Y.K.; Uribe-Romo, F.J.; Kim, J.; O'Keeffe, M.; Yaghi, O.M. Isoreticular Expansion of Metal-Organic Frameworks with Triangular and Square Building Units and the Lowest Calculated Density for Porous Crystals. *Inorg. Chem.* **2011**, *50*, 9147–9152. [CrossRef] [PubMed]

45. Forgan, R.S. Modulated self-assembly of metal–organic frameworks. *Chem. Sci.* **2020**, *11*, 4546–4562. [CrossRef] [PubMed]

46. Wang, C.; Zhou, M.; Ma, Y.; Tan, H.; Wang, Y.; Li, Y. Hybridized Polyoxometalate-Based Metal–Organic Framework with Ketjenblack for the Nonenzymatic Detection of H2O2. *Chem. Asian. J.* **2018**, *13*, 2054–2059. [CrossRef]

47. Ramaraghavulu, R.; Rao, V.K.; Devarayapalli, K.C.; Yoo, K.; Nagajyothi, P.C.; Shim, J. Green synthesized AgNPs decorated on Ketjen black for enhanced catalytic dye degradation. *Res. Chem. Intermed.* **2021**, *47*, 637–648. [CrossRef]

48. Li, K.; Liu, Q.; Cheng, H.; Hu, M.; Zhang, S. Classification and carbon structural transformation from anthracite to natural coaly graphite by XRD, Raman spectroscopy, and HRTEM. *Spectrochim. Acta Part A* **2021**, *249*, 119286. [CrossRef] [PubMed]

49. Meier, H.; Bienz, S.; Bigler, L.; Fox, T. *Spektroskopische Methoden in der Organischen Chemie*, 8th ed.; Georg Thieme: Stuttgart, Germany, 2011.

50. Wu, Y.; Song, X.; Li, S.; Zhang, J.; Yang, X.; Shen, P.; Gao, L.; Wei, R.; Zhang, J.; Xiao, G. 3D-monoclinic M–BTC MOF (M = Mn, Co, Ni) as highly efficient catalysts for chemical fixation of CO_2 into cyclic carbonates. *J. Ind. Eng. Chem.* **2018**, *58*, 296–303. [CrossRef]

51. Maruthapandian, V.; Kumaraguru, S.; Mohan, S.; Saraswathy, V.; Muralidharan, S. An Insight on the Electrocatalytic Mechanistic Study of Pristine Ni MOF (BTC) in Alkaline Medium for Enhanced OER and UOR. *ChemElectroChem* **2018**, *5*, 2795–2807. [CrossRef]

52. Yaghi, O.M.; Li, H.; Groy, T.L. Construction of Porous Solids from Hydrogen-Bonded Metal Complexes of 1,3,5-Benzenetricarboxylic Acid. *J. Am. Chem. Soc.* **1996**, *118*, 9096–9101. [CrossRef]

53. Vuong, G.-T.; Pham, M.-H.; Do, T.-O. Direct synthesis and mechanism of the formation of mixed metal Fe_2Ni-MIL-88B. *CrystEngComm* **2013**, *15*, 9694–9703. [CrossRef]

54. Vu, H.T.; Nguyen, M.B.; Vu, T.M.; Le, G.H.; Pham, T.T.T.; Nguyen, T.D.; Vu, T.A. Synthesis and Application of Novel Nano Fe-BTC/GO Composites as Highly Efficient Photocatalysts in the Dye Degradation. *Top. Catal.* **2020**, *63*, 1046–1055. [CrossRef]

55. He, J.; Lu, X.; Yu, J.; Wang, L.; Song, Y. Hierarchical $Co(OH)_2$ nanostructures/glassy carbon electrode derived from Co(BTC) metal–organic frameworks for glucose sensing. *J. Nanopart. Res.* **2016**, *18*, 184. [CrossRef]

56. Gan, Q.; He, H.; Zhao, K.; He, Z.; Liu, S. Morphology-dependent electrochemical performance of Ni-1,3,5-benzenetricarboxylate metal-organic frameworks as an anode material for Li-ion batteries. *J. Colloid Interface Sci.* **2018**, *530*, 127–136. [CrossRef] [PubMed]

57. Wang, L.; Ren, L.; Wang, X.; Feng, X.; Zhou, J.; Wang, B. Multivariate MOF-Templated Pomegranate-Like Ni/C as Efficient Bifunctional Electrocatalyst for Hydrogen Evolution and Urea Oxidation. *ACS Appl. Mater Interfaces* **2018**, *10*, 4750–4756. [CrossRef]

58. Thommes, M.; Kaneko, K.; Neimark, A.V.; Olivier, J.P.; Rodriguez-Reinoso, F.; Rouquerol, J.; Sing, K.S.W. Physisorption of gases, with special referenceto the evaluation of surface area and pore size distribution (IUPAC Technical Report). *Pure Appl. Chem.* **2015**, *87*, 1051–1069. [CrossRef]

59. Wade, C.R.; Dincă, M. Investigation of the synthesis, activation, and isosteric heats of CO_2 adsorption of the isostructural series of metal–organic frameworks $M_3(BTC)_2$ (M = Cr, Fe, Ni, Cu, Mo, Ru). *Dalton Trans.* **2012**, *41*, 7931–7938. [CrossRef]

60. Israr, F.; Chun, D.; Kim, Y.; Kim, D.K. High yield synthesis of Ni-BTC metal–organic framework with ultrasonic irradiation: Role of polar aprotic DMF solvent. *Ultrason. Sonochem.* **2016**, *31*, 93–101. [CrossRef]

61. Israr, F.; Kim, D.K.; Kim, Y.; Chun, W. Scope of various solvents and their effects on solvothermal synthesis of Ni-BTC. *Quim. Nova* **2016**, *39*, 669–675. [CrossRef]

62. Long, X.; Ma, Z.; Yu, H.; Gao, X.; Pan, X.; Chen, X.; Yang, S.; Yi, Z. Porous FeNi oxide nanosheets as advanced electrochemical catalysts for sustained water oxidation. *J. Mater. Chem. A* **2016**, *4*, 14939–14943. [CrossRef]
63. Wang, Q.; Wei, C.; Li, D.; Guo, W.; Zhong, D.; Zhao, Q. FeNi-based bimetallic MIL-101 directly applicable as an efficient electrocatalyst for oxygen evolution reaction. *Microporous Mesoporous Mater.* **2019**, *286*, 92–97. [CrossRef]
64. Louie, M.W.; Bell, A.T. An Investigation of Thin-Film Ni–Fe Oxide Catalysts for the Electrochemical Evolution of Oxygen. *J. Am. Chem. Soc.* **2013**, *135*, 12329–12337. [CrossRef] [PubMed]
65. Bates, M.K.; Jia, Q.; Doan, H.; Liang, W.; Mukerjee, S. Charge-Transfer Effects in Ni–Fe and Ni–Fe–Co Mixed-Metal Oxides for the Alkaline Oxygen Evolution Reaction. *ACS Catal.* **2016**, *6*, 155–161. [CrossRef]
66. Corrigan, D.A. The Catalysis of the Oxygen Evolution Reaction by Iron Impurities in Thin Film Nickel Oxide Electrodes. *J. Electrochem. Soc.* **1987**, *134*, 377–384. [CrossRef]
67. McCrory, C.C.L.; Jung, S.; Peters, J.C.; Jaramillo, T.F. Benchmarking Heterogeneous Electrocatalysts for the Oxygen Evolution Reaction. *J. Am. Chem. Soc.* **2013**, *135*, 16977–16987. [CrossRef]
68. McCrory, C.C.L.; Jung, S.; Ferrer, I.M.; Chatman, S.M.; Peters, J.C.; Jaramillo, T.F. Benchmarking Hydrogen Evolving Reaction and Oxygen Evolving Reaction Electrocatalysts for Solar Water Splitting Devices. *J. Am. Chem. Soc.* **2015**, *137*, 4347–4357. [CrossRef]
69. Lee, M.; Oh, H.-S.; Cho, M.K.; Ahn, J.-P.; Hwang, Y.J.; Min, B.K. Activation of a Ni electrocatalyst through spontaneous transformation of nickel sulfide to nickel hydroxide in an oxygen evolution reaction. *Appl. Catal. B* **2018**, *233*, 130–135. [CrossRef]
70. Möller, S.; Barwe, S.; Masa, J.; Wintrich, D.; Seisel, S.; Baltruschat, H.; Schuhmann, W. Online Monitoring of Electrochemical Carbon Corrosion in Alkaline Electrolytes by Differential Electrochemical Mass Spectrometry. *Angew. Chem. Int. Ed.* **2020**, *59*, 1585–1589. [CrossRef]
71. Ji, S.G.; Kim, H.; Lee, W.H.; Oh, H.-S.; Choi, C.H. Real-time monitoring of electrochemical carbon corrosion in alkaline media. *J. Mater. Chem. A* **2021**, *9*, 19834–19839. [CrossRef]
72. Li, G.; Anderson, L.; Chen, Y.; Pan, M.; Chuang, P.-Y.A. New insights into evaluation catalyst activity and stability for oxygen evolution reactions in alkaline media. *Sustain. Energy Fuels* **2018**, *2*, 237–251. [CrossRef]
73. Krasilshchikov, A.I. Intermediate stages of oxygen anodic evolution. *Zhurnal Fiz. Khim.* **1963**, *37*, 531–537.
74. Yuan, Y.F.; Xia, X.H.; Wu, J.B.; Yang, J.L.; Chen, Y.B.; Guo, S.Y. Nickel foam-supported porous Ni(OH)$_2$/NiOOH composite film as advanced pseudocapacitor material. *Electrochim. Acta* **2011**, *56*, 2627–2632. [CrossRef]
75. Hall, D.S.; Lockwood, D.J.; Bock, C.; MacDougall, B.R. Nickel hydroxides and related materials: A review of their structures, synthesis and properties. *Proc. R. Soc. A* **2015**, *471*, 20140792. [CrossRef] [PubMed]
76. Huang, J.; Chen, J.; Yao, T.; He, J.; Jiang, S.; Sun, Z.; Liu, Q.; Cheng, W.; Hu, F.; Jiang, Y.; et al. CoOOH Nanosheets with High Mass Activity for Water Oxidation. *Angew. Chem.* **2015**, *54*, 8846–8851. [CrossRef]
77. Hu, J.; Li, S.; Chu, J.; Niu, S.; Wang, J.; Du, Y.; Li, Z.; Han, X.; Xu, P. Understanding the Phase-Induced Electrocatalytic Oxygen Evolution Reaction Activity on FeOOH Nanostructures. *ACS Catal.* **2019**, *9*, 10705–10711. [CrossRef]
78. Sivanantham, A.; Ganesan, P.; Vinu, A.; Shanmugam, S. Surface Activation and Reconstruction of Non-Oxide-Based Catalysts Through in Situ Electrochemical Tuning for Oxygen Evolution Reactions in Alkaline Media. *ACS Catal.* **2020**, *10*, 463–493. [CrossRef]
79. Zhu, J.; Zhou, G.; Ding, Y.; Wang, Z.; Hu, Y.; Zou, M. A Facile Route to Oriented Nickel Hydroxide Nanocolumns and Porous Nickel Oxide. *J. Phys. Chem. C* **2007**, *111*, 5622–5627. [CrossRef]
80. Zheng, W.; Liu, M.; Lee, L.Y.S. Electrochemical Instability of Metal–Organic Frameworks: In Situ Spectroelectrochemical Investigation of the Real Active Sites. *ACS. Catal.* **2020**, *10*, 81–92. [CrossRef]
81. Napporn, T.W.; Holade, Y.; Kokoh, B.; Mitsushima, S.; Mayer, K.; Eichberger, B.; Hacker, V. Electrochemical Measurement Methods and Characterization on the Cell Level. In *Fuel Cells and Hydrogen: From Fundamentals to Applied Research*; Hacker, V., Mitsushima, S., Eds.; Elsevier, B.V.: Amsterdam, The Netherlands, 2018; pp. 175–214. [CrossRef]
82. Pretsch, E.; Bühlmann, P.; Badertscher, M. *Spektroskopische Daten Zur Strukturaufklärung Organischer Verbindungen*, 5th ed., Springer: Berlin/Heidelberg, Germany, 2010.
83. Young, K.-H.; Wang, L.; Yan, S.; Liao, X.; Meng, T.; Shen, H.; Mays, W.C. Fabrications of High-Capacity Alpha-Ni(OH)2. *Batteries* **2017**, *3*, 6. [CrossRef]
84. Lu, C.-T.; Chiu, Y.-W.; Li, M.-J.; Hsueh, K.-L.; Hung, J.-S. Reduction of the Electrode Overpotential of the Oxygen Evolution Reaction by Electrode Surface Modification. *Int. J. Electrochem.* **2017**, *2017*, 7494571. [CrossRef]

 *molecules*

Article

The Facile Deposition of Pt Nanoparticles on Reduced Graphite Oxide in Tunable Aryl Alkyl Ionic Liquids for ORR Catalysts

Dennis Woitassek [1], Swantje Lerch [2], Wulv Jiang [3], Meital Shviro [3,*], Stefan Roitsch [4], Thomas Strassner [2,*] and Christoph Janiak [1,*]

[1] Institut für Anorganische Chemie und Strukturchemie, Heinrich-Heine-Universität Düsseldorf, 40225 Düsseldorf, Germany; woitasse@hhu.de
[2] Physikalische Organische Chemie, Technische Universität Dresden, 01062 Dresden, Germany; swantje.lerch@tu-dresden.de
[3] Forschungszentrum Jülich GmbH, Institute of Energy and Climate Research, IEK-14: Electrochemical Process Engineering, 52425 Jülich, Germany; j.wulv@fz-juelich.de
[4] Institut für Physikalische Chemie, Universität zu Köln, 50939 Köln, Germany; sroitsch@uni-koeln.de
* Correspondence: m.shviro@fz-juelich.de (M.S.); thomas.strassner@tu-dresden.de (T.S.); janiak@uni-duesseldorf.de (C.J.); Tel.: +49-2118112286 (C.J.)

Abstract: In this study, we present the facile formation of platinum nanoparticles (Pt-NPs) on reduced graphite oxide (rGO) (Pt-NP@rGO) by microwave-induced heating of the organometallic precursor ((MeCp)PtMe$_3$ in different tunable aryl alkyl ionic liquids (TAAIL). In the absence of rGO, transmission electron microscopy (TEM) reveals the formation of dense aggregates of Pt-NPs, with primary particle sizes of 2 to 6 nm. In contrast, in the Pt-NP@rGO samples, Pt-NPs are homogeneously distributed on the rGO, without any aggregation. Pt-NP@rGO samples are used as electrode materials for oxygen reduction reaction (ORR), which was assessed by cyclic voltammetry (CV) and linear sweep voltammetry (LSV). The electrochemical surface area (ECSA) and mass-specific activity (MA) increase up to twofold, compared with standard Pt/C 60%, making Pt-NP@rGO a competitive material for ORR.

Keywords: platinum; nanoparticles; electrochemistry; oxygen reduction reaction; reduced graphite oxide; microwave; ionic liquid; tunable aryl alkyl ionic liquid

---

Citation: Woitassek, D.; Lerch, S.; Jiang, W.; Shviro, M.; Roitsch, S.; Strassner, T.; Janiak, C. The Facile Deposition of Pt Nanoparticles on Reduced Graphite Oxide in Tunable Aryl Alkyl Ionic Liquids for ORR Catalysts. *Molecules* **2022**, *27*, 1018. https://doi.org/10.3390/molecules 27031018

Academic Editors: Jingqi Guan and Yin Wang

Received: 17 December 2021
Accepted: 27 January 2022
Published: 2 February 2022

Publisher's Note: MDPI stays neutral with regard to jurisdictional claims in published maps and institutional affiliations.

1. Introduction

Platinum nanoparticles (Pt-NPs) are important catalysts and are often used as benchmark materials in the field of electrochemistry for the oxygen reduction reaction (ORR) [1–5], the hydrogen evolution reaction (HER) [6,7], and the methanol oxidation reaction (MOR) [8–10]. In general, the stability, catalytic activity, and chemical selectivity of Pt-NPs depend strongly on their size, shape, alloy composition, surface structure, and surface accessibility [2,5,11]. It is, therefore, important to control these parameters via the chosen synthetic method and, if possible, influence multiple parameters at the same time during NP synthesis [2,5,11].

Smaller Pt-NP sizes show higher metal-mass-based catalytic activity than larger particles or bulk-material [5,11,12]. At the same time, a crucial issue for NPs is their tendency to coalescence, which can occur as agglomeration or Ostwald ripening, increasing their size and reducing the catalytic activity, thus requiring stabilizers [13–15].

We have recently reported the wet-chemical synthesis of Ru- and Ir-NPs in tunable aryl alkyl ionic liquids (TAAILs) based on the 1-aryl-3-alkyl-substituted imidazolium motif [16]. Ionic liquids (ILs) are salts with a melting point below 100 °C and can be used as solvents and as stabilizing agents for M-NPs, due to their high ionic charge, polarity, dielectric constant, and electrostatic and steric interaction with the M-NPs [13–15,17,18]. All used TAAILs were suitable for the formation of very small (<5 nm) NPs, which were stable over months in the IL medium. An interesting observation was that the nanoparticle separation and aggregation varied strongly as a function of the aryl substituent of the TAAILs.

In addition, Supporting Materials can stabilize M-NPs, allowing for easier handling of the M-NPs and providing for additional tuning options for applications. Carbon materials as support for Pt-NPs are commonly used in the field of electrochemistry because of their high electrical conductivity, low weight, low cost, easy production, and manifold structures [6,19]. Our group has already shown the usage of (thermally) reduced graphite oxide (rGO, also known as TRGO) as M-NP support [20,21], which also offers remarkable properties as support for electrode materials [22–24]. rGO increases the electrochemical activity of Pt-NPs [19], stabilizes very small M-NPs [25,26], and offers improved protection for supported NPs against poisoning, compared with available standard carbon black materials [20]. Since rGO provides less functional groups that can interact with M-NPs than graphite oxide, the deposition of Pt-NPs on rGO (Pt-NP@rGO) in ILs remains difficult and was only successful after additional rGO functionalization with thiol groups [20] to improve the interaction between Pt-NPs and its support. Alternatively, porous organic polymers with donor atoms such as covalent triazine frameworks (CTFs) were successfully used as support [27].

In this article, we present a simple and facile way to deposit Pt-NPs in situ on rGO during a microwave-assisted reaction in ionic liquids. Novel imidazolium-ILs, which are distinguished aryl and alkyl substitutions, are used as reaction media as well as stabilizing agents. We also show that the obtained Pt-NP@rGO are electrochemically active as heterogeneous catalysts for ORR, a vital reaction for proton exchange membrane fuel cells (PEMFC).

2. Results

2.1. rGO and TAAIL Presentation

rGO was synthesized in a two-step oxidation–thermal reduction process according to procedures by Hummers [28], as described in more detail in our previous studies [20,21], with the thermal reduction occurring at 400 °C. CHNS elemental analysis showed a weight percentage (wt%) of 80% carbon, which indicates that functional oxygen-containing groups still remain on the surface of rGO [21]. The exact values can be found in the Supporting Information (SI), Table S2.

The synthesis and characterization of the TAAILs used here were described elsewhere [15,29,30]. TAAILs refer to 1-n-alkyl-3-arylimidazolium ILs (Scheme 1). The length of the n-alkyl chain varies here between 4 and 11 carbon atoms, and the aryl groups were 4-methoxy-, 2,4-dimethyl, 4-bromo- and 2-methylphenyl.

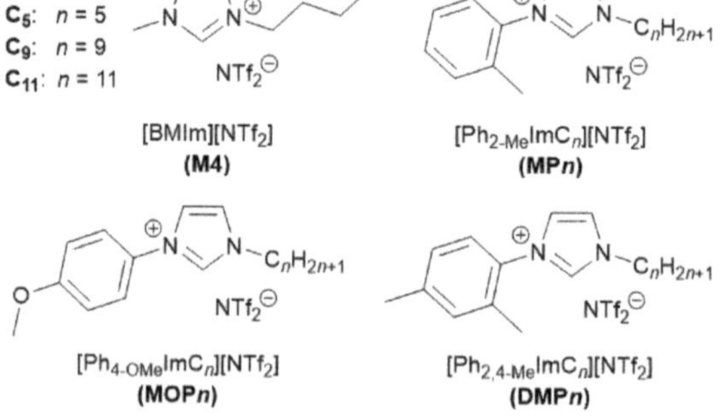

Scheme 1. Structures of utilized tunable aryl alkyl ionic liquids (TAAIL) with 1,3 disubstituted imidazolium cations and the bis(trifluoromethylsulfonyl)imide anion. The abbreviation is derived from the phenyl substitution and the alkyl chain length.

The TAAILs can be obtained on a large gram scale, and the synthesis allows manifold variations of the aryl and alkyl groups [15,29,30]. The anion purity of the TAAILs was characterized via ion chromatography and has always been found above at least 97% and more often 99%. Thermogravimetric analysis (TGA) of the TAAILs showed decomposition between 400 °C and 426 °C, which allows them to be used in the Pt-NP-forming microwave reactions at ~200 °C.

2.2. Pt-NP Synthesis and Characterization

Following previous research [13], (η^5-methylcyclopentadienyl)trimethylplatinum(IV) ((MeCp)PtMe$_3$) was used as a Pt precursor, which can be decomposed in various ways (thermal, photolytic, sonolytic) at relatively mild conditions and without any additives [13,20]. The decomposition of the precursor produces volatile side products, which do not contaminate the NP surface [31–33].

The general synthetic procedure of the Pt-NPs is depicted in Scheme 2. Fixed amounts of Pt precursor and rGO were suspended in the IL to achieve 1 wt% or 2 wt% Pt-NP and 0.5 wt% or 1 wt% rGO in the IL dispersion. After stirring overnight and for 3 h of sonication, the dispersion was placed in a microwave reactor, heated at 40 W to 200 °C, and kept at this temperature for 10 min. Black Pt-NP@rGO dispersions were reproducibly obtained.

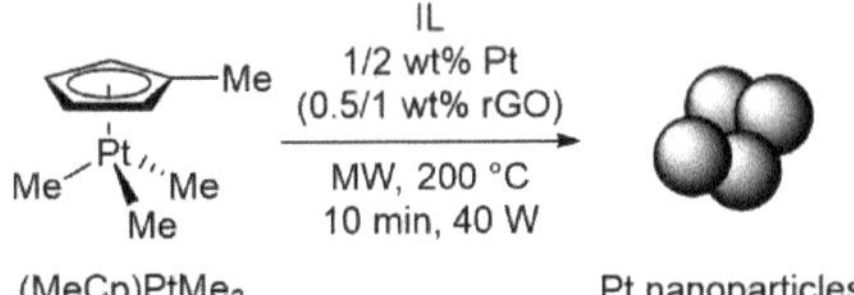

Scheme 2. Reaction conditions for the synthesis of platinum nanoparticles (Pt-NPs) via microwave-induced heating in ionic liquids (ILs). The mass of the Pt precursor was set to achieve 1 wt% or 2 wt% Pt-NP in the IL dispersion upon complete decomposition (0.5 wt% or 1 wt% reduced graphite oxide (rGO) used for Pt-NPs stabilized on rGO (Pt-NP@rGO) in brackets).

Selected samples and their respective particle size determined via transmission electron microscopy (TEM) are shown in Table 1. In the Supporting Information, Table S1, a full list of prepared samples is given. The nanocrystallinity of the Pt particles was verified by powder X-ray diffraction (PXRD) (Figure 1 and Figures S2–S6 in the Supplementary Information). The Pt-NPs form in the face-centered cubic (fcc) structure, typical for crystalline Pt. The comparison of the crystallite sizes of the NPs calculated with the Scherrer equation (Section 4.1, Equation (1)) and the Pt-NP sizes determined via TEM showed a very good agreement (Table S1, Supplementary Information). Furthermore, thermogravimetric analysis (TGA) of elected samples (Figure S28 in the Supplementary Information) gives a residual Pt mass between 41 wt% and 54 wt% or a reciprocal combined rGO and IL content between 46 wt% and 59 wt%. The mass-temperature profile in TGA did not allow a distinction between rGO and IL. From an estimate of the IL content by CHNS analysis (Table S2), the rGO mass was then approximated between 40 wt% and 54 wt% (Table S4).

Similar to Ir-NPs, Ru-NPs [15], and Pt-NPs [13], the microwave-assisted heating of the platinum metal precursor (MeCp)PtMe$_3$ in ionic liquid dispersion results in small M-NP sizes, largely between 2 nm and 5 nm. The efficient energy uptake of ILs enables fast heating and results in rapid decomposition of the Pt precursor with a high nucleation rate of Pt-NPs, which themselves then absorb microwave radiation. This leads to "hot spots", with a further increase in localized temperature. Small particle sizes can only be achieved if the TAAIL is able to stabilize the newly formed NPs from the very beginning of the growth process. The Pt-NPs show similar sizes mostly independent of the choice of TAAIL and the wt% Pt to IL (Table S1, Supplementary Information). An exception is the TAAIL with the 4-methoxyphenyl substituent (MOP5 to MOP11), where also larger particles, up to 6 nm, were obtained.

Table 1. Summary of Pt-NPs and Pt-NP@rGO in tunable aryl alkyl ionic liquids (TAAILs). Further information can be found in Table S1 in the Supplementary Information.

Sample Name X/Y = wt% Pt/rGO	TAAIL Used	wt% Pt [1] (X/-)	wt% rGO [1] (-/Y)	Average Particle Size [2] [nm]
M4-1	[BMIm][NTf$_2$]	1	-	1
M4_rGO-1/1	[BMIm][NTf$_2$]	1	1	2 ± 1
MP4-1	[Ph$_{2\text{-Me}}$ImC$_4$][NTf$_2$]	1	-	5 ± 1
MP4_rGO-1/1	[Ph$_{2\text{-Me}}$ImC$_4$][NTf$_2$]	1	1	3 ± 1
MP5-1	[Ph$_{2\text{-Me}}$ImC$_5$][NTf$_2$]	1	-	4 ± 1
MP5_rGO-2/1	[Ph$_{2\text{-Me}}$ImC$_5$][NTf$_2$]	2	1	2 ± 1
MP9-1	[Ph$_{2\text{-Me}}$ImC$_9$][NTf$_2$]	1	-	4 ± 1
MP9_rGO-1/0.5	[Ph$_{2\text{-Me}}$ImC$_9$][NTf$_2$]	1	0.5	3 ± 1
DMP9-1	[Ph$_{2,4\text{-Me}}$ImC$_9$][NTf$_2$]	1	-	3 ± 1
DMP9_rGO-1/1	[Ph$_{2,4\text{-Me}}$ImC$_9$][NTf$_2$]	1	1	6 ± 1
MOP5-1	[Ph$_{4\text{-OMe}}$ImC$_5$][NTf$_2$]	1	-	2 ± 1
MOP5_rGO-1/1	[Ph$_{4\text{-OMe}}$ImC$_5$][NTf$_2$]	1	1	5 ± 1

[1] wt% in TAAIL dispersion. [2] Average particle sizes obtained from transmission electron microscopy (TEM) measurements. At least 200 particles were used for size determination. See Materials and Methods for more information.

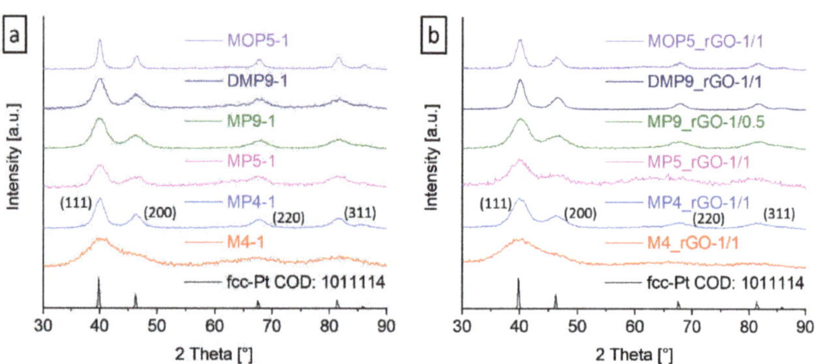

Figure 1. Powder X-ray diffraction (PXRD) patterns of synthesized Pt-NPs (**a**) and Pt-NP@rGO (**b**) in ILs, together with the simulation for fcc-Pt and its indexed reflections (Crystallographic open database fcc-Pt: 1011114). The -X and -X/Y numbers refer to wt% Pt (X) and wt% rGO (Y) in samples. Additional PXRD patterns can be found in Figures S2–S6 in the Supporting Information (SI).

2.3. Pt-NPs Aggregation in TAAILs

As seen before for Ru- and Ir-NPs, [15], the TAAILs allow the microwave-induced synthesis of Pt-NPs, but strong Pt-NP aggregation was observed (in the absence of rGO) for all TAAILs, independent of the alkyl chain length or the aryl substitution (Supplementary Information, Figures S7–S27), whereas the comparative IL [BMIm][NTf$_2$] gave well-separated NPs (Figure 2 top, Figure S7). Different from the synthesis of Ru- and Ir-NPs, [15] even the 4-methoxyphenyl- and 2,4-dimethylphenyl-TAAIL did not prevent strong aggregation.

Aggregation, however, could be prevented upon deposition on rGO by performing the (MeCp)PtMe$_3$ precursor decomposition in the presence of rGO. Representative TEM images of Pt-NPs synthesized in different TAAILs without and with rGO are shown in Figures 2 and 3. The rGO-free samples consist of highly aggregated Pt-NPs, with large, dense structures, with dimensions of several tens of nanometers. The degree of aggregation for the Pt-NPs appears independent of the TAAIL.

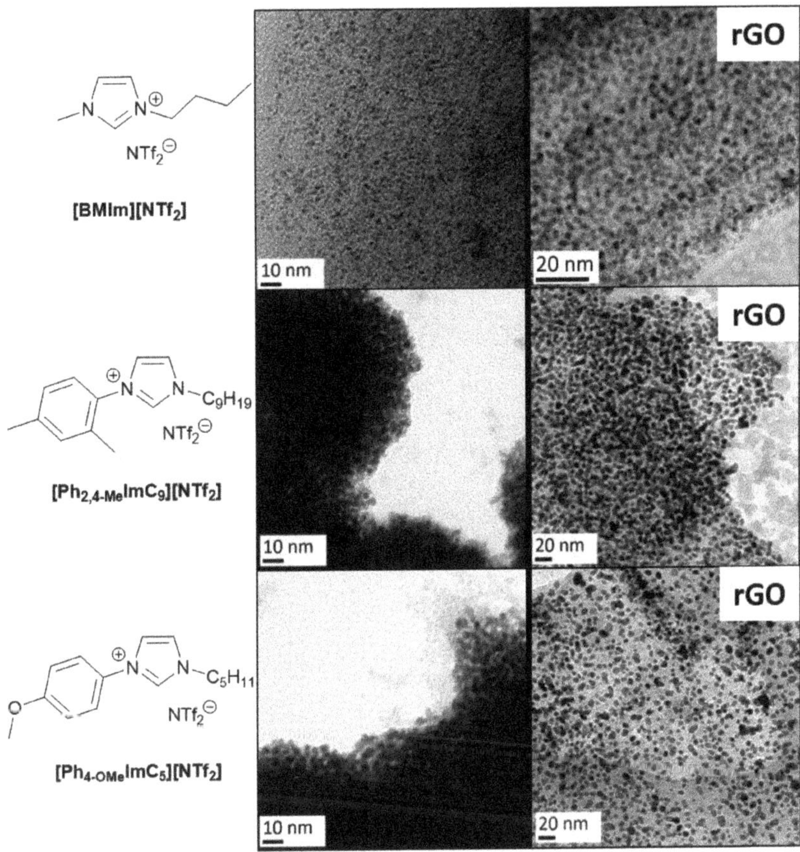

Figure 2. TEM images of Pt-NPs and Pt-NP@rGO composites obtained in the IL (from top to bottom) [BMIm][NTf$_2$] (**M4-1, M4_rGO-1/1**) and the TAAILs [Ph$_{2,4\text{-Me}}$ImC$_9$][NTf$_2$] (**DMP9-1, DMP9_rGO-1/1**) and [Ph$_{4\text{-OMe}}$ImC$_5$][NTf$_2$] (**MOP5-1, MOP5_rGO-1/1**). TEM images for Pt-NPs are shown on the left and for Pt-NP@rGO composites on the right. The rGO sheets on which the Pt-NP are deposited are clearly seen by their wrinkles at lower magnification (e.g., Figures S10, S12, S16, S17, S19, S25, and S27 in the Supplementary Information).

In previous research, we discussed the necessity of thiol functional groups on rGO to deposit Pt nanoparticles as Pt-NP@rGO when using the IL [BMIm][BF$_4$] [20]. In contrast, with the TAAILs used in this study (and also with the [NTf$_2$]-IL M4 (Figure S8 in the Supplementary Information)), Pt-NPs can be effectively stabilized on rGO without thiol functional groups, resulting in a rather homogenous layer of Pt-NPs on the surface of the rGO (Figures 2 and 3). Compared with the rGO-free samples, nearly no additional agglomeration of NPs can be found, which indicates that the NPs are effectively anchored to prevent aggregation.

The stability of prepared Pt-NPs and Pt-NP@rGO samples was followed by storing them dried under air at room temperature for at least 6 months. PXRD patterns and the calculated crystallite sizes (Equation (1)) from representative samples are given in Table S3 and Figures S3–S6 in the Supplementary Information. Additionally, a TEM image of an rGO-free and a Pt-NP@rGO sample after 1 year did not show significant particle growth or changes in aggregation (Figures S26 and S27 in Supplementary Information).

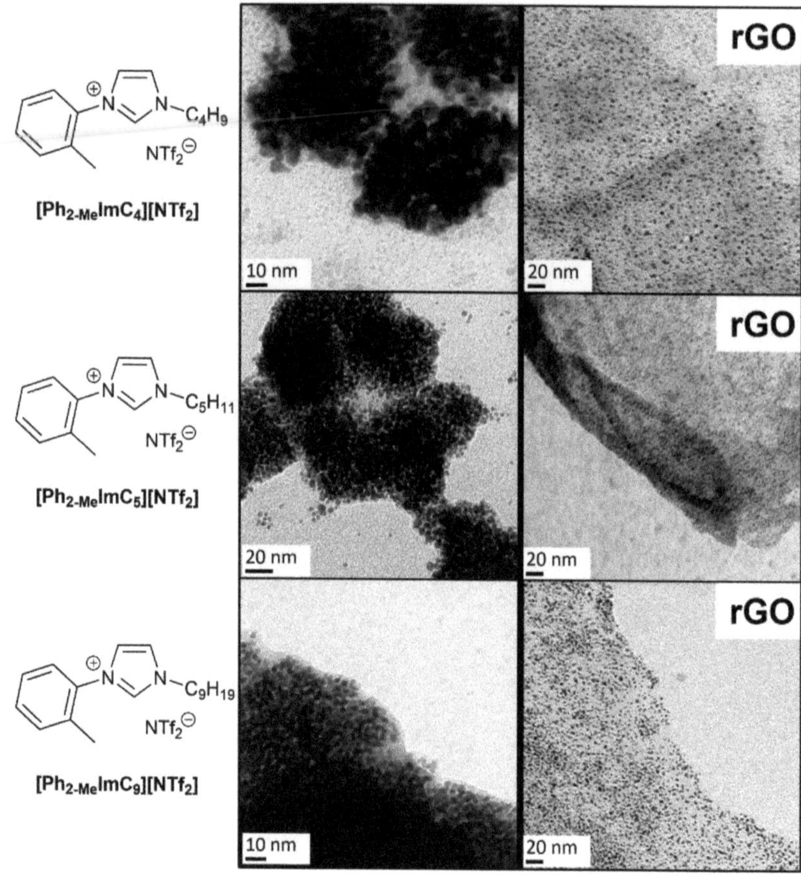

Figure 3. TEM images of Pt-NPs and Pt-NP@rGO composites obtained in the TAAILs (from top to bottom): [Ph2-MeImC4][NTf2] (**MP4-1, MP4_rGO-1/1**); [Ph2-MeImC5][NTf2] (**MP5-1, MP5_rGO-2/1**); [Ph2-MeImC9][NTf2] (**MP9-1, MP9_rGO-1/0.5**). TEM images for Pt-NPs are shown on the left and for Pt-NP@rGO composites on the right. The rGO sheets on which the Pt-NP are deposited are clearly seen by their wrinkles at lower magnification (e.g., Figures S10, S12, S16, S17, S19, S25, and S27 in the Supplementary Information).

2.4. Electrochemical Catalysis

The electrochemical performance of obtained Pt-NP@rGO for the ORR was examined with commercial Pt/C 60% as internal benchmark material in comparison, to reduce deviations by the use of an external standard. As samples, **MP4_rGO-1/1**, **MP5_rGO-2/1**, **MP9_rGO-1/0.5**, and **MOP5_rGO-1/1** were chosen because of their low size and homogeneity, as depicted in Figure 4. The cyclic voltammogram (CV) curves were recorded in N_2-saturated 0.1 M $HClO_4$, with a sweep rate of 20 mV s^{-1} and, after 50 activation cycles, against reversible hydrogen electrode (RHE, Figure 4a). Afterward, ORR polarization curves (linear sweep voltammetry (LSV)) with a sweep rate of 10 mV s^{-1} were recorded in O_2-saturated 0.1 M $HClO_4$. These measurements were repeated three times each, and the respective combined average CVs and LSVs are shown in Figure 4a,b. The electrochemical surface area (ECSA) was determined by integrating the hydrogen underpotential deposition (H_{upd}) charge (H_{upd} charge to surface area conversion constant: 210 µC cm^{-2}) in the obtained CVs, and, with the obtained LSVs, mass-specific activities (MAs) at 0.9 V were calculated. The obtained ECSA and MA are compared in Table 2 to the results of Pt-NPs deposited on rGO via different reaction methods from previous studies of Badam et al. [34], Daş et al. [35], and Teran-Salgadon et al. [36].

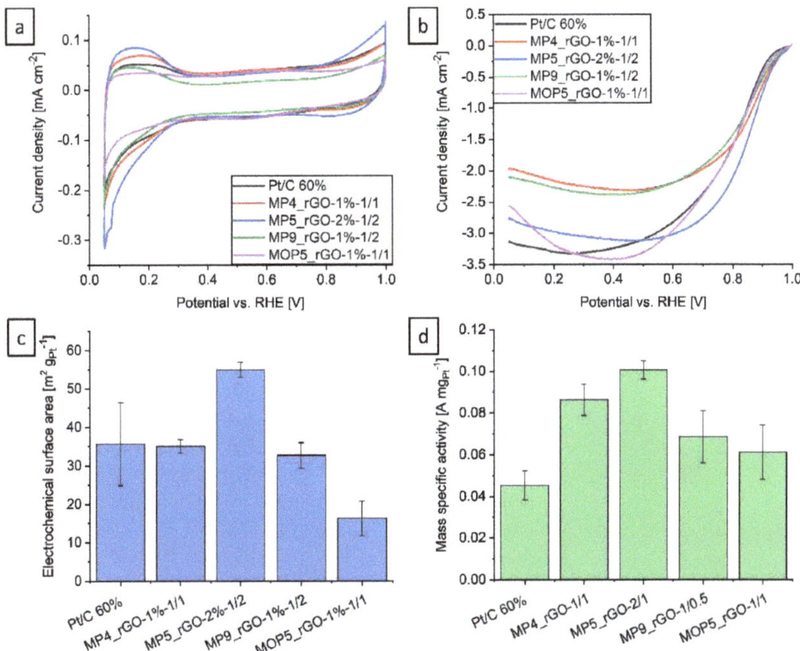

Figure 4. Results obtained via electrochemical measurements: (**a**) averaged cyclic voltammetry (CV) in N_2-atmosphere and at 20 mV s^{-1} sweep rate of the four samples **MP4_rGO-1/1**, **MP5_rGO-2/1**, **MP9_rGO-1/0.5**, **MOP5_rGO-1/1**, and a Pt/C 60% standard material; (**b**) averaged oxygen reduction reaction (ORR) polarization curves (linear sweep voltammetry (LSV)) of the same samples in O_2-saturated atmosphere at 10 mV s^{-1} sweep rate. From these data, (**c**) the electrochemical surface area (ECSA) and (**d**) the mass-specific activity (MA) at 0.9 V were calculated.

Table 2. Electrochemical surface area (ECSA) and mass-specific activity (MA) of Pt-NP@rGO synthesized with and without IL.

Sample Name	IL Used	ECSA [1] [m^2 g$_{Pt}^{-1}$]	MA [1] [mA mg$_{Pt}^{-1}$]
60% Pt/C	-	36 ± 11	45 ± 7
MP4_rGO-1/1	[Ph$_{2\text{-Me}}$ImC$_4$][NTf$_2$]	35 ± 2	86 ± 7
MP5_rGO-2/1	[Ph$_{2\text{-Me}}$ImC$_4$][NTf$_2$]	55 ± 2	101 ± 4
MP9_rGO-1/0.5	[Ph$_{2\text{-Me}}$ImC$_4$][NTf$_2$]	33 ± 3	68 ± 12
MOP5_rGO-1/1	[Ph$_{4\text{-OMe}}$ImC$_4$][NTf$_2$]	16 ± 5	61 ± 13
Pt-TMIM-rGO [34]	[C$_{12}$ImC$_1$][GO]	56.8	346
Pt/rGO (DMF) [35]	-	28.1	26
Pt/rGO (HYD) [35]	-	18.2	7
Pt/rGO [36]	-	14.0	92

[1] Standard deviation obtained by three measurements (see Materials and Methods, Section 4.3).

All CVs of the materials, shown in Figure 4a, have well-pronounced underpotential hydrogen deposition–desorption areas between 0.05 V and 0.40 V, with **MP4_rGO-1/1** and **MP5_rGO-2/1** having a larger area than the commercial Pt/C catalyst. No contaminations or side reactions are recognizable in the CVs. The calculated ECSA, shown in Figure 4c, reveals the highest value for **MP5_rGO-2/1**, with a 1.5-fold increase, compared with Pt/C 60%, while the other samples show similar values as the benchmark material. This shows that rGO can be effectively used as in situ carbon source, in addition to the most commonly used Vulcan XC-72R, which is only added for preparing the catalyst ink. The ORR polarization curves (Figure 4b) of the samples show better onset potentials than the benchmark material and different current density plateaus between 0.20 V and 0.60 V. The presented materials reach their respective plateau at higher voltages but at lower currents than Pt/C 60%. From the LSVs the mass-specific activity (MA) is calculated (Figure 4d). The highest MA is achieved by **MP5_rGO-2/1**, with a promising twofold increase, compared with the Pt/C 60% benchmark, while **MP4_rGO-1/1**, **MP9_rGO-1/0.5**, and **MOP5_rGO-1/1** also show an increase of more than 1.8-fold, 1.4-fold, and 1.25-fold, respectively, demonstrating a better performance of Pt-NP@rGO for ORR than Pt/C 60%. Calculated Tafel plots of the samples, shown in Figure S29 in the Supplementary Information, are similar to the Tafel plot of Pt/C 60% and indicate a four-electron pathway, with the adsorption of oxygen as a rate-limiting step [37]. The Koutecký–Levich plot calculated from LSVs of **MP4_rGO-1/1** at different rotation speeds (Figures S30 and S31 in the Supplementary Information) also shows a dominant four-electron pathway toward the ORR process.

A comparison of previous research on Pt-NP@rGO shows the highest ECSA and MA for Pt-NP decorated on tetradecyl-methyl-imidazolium ionic liquid-treated graphene (Pt-TMIm-rGO) by Badam et al. [34], albeit with a similar ECSA to the 55 m^2 g_{Pt}^{-1} of **MP5_rGO-2/1**. Notably, the **MP_rGO** samples have higher ECSA values, compared with the results of the Pt/rGO probes of Daş et al. [35] and Teran-Salgadon et al. [36], much higher MA values than obtained by Daş et al. [35] and similar to slightly smaller MA values than given by Teran-Salgadon et al. [36]. From the comparison, it becomes evident that ILs can significantly increase the catalytic activity of Pt-NP@rGO samples.

2.5. Stability Testing

MP4_rGO-1/1, the most active sample for ORR, was further analyzed with an electrochemical stability test. After the material was activated via CV (20 cycles), 5k cycles between 0.5 and 1.0 V_{RHE} were carried out. The obtained LSVs and calculated MA after activation and after 5k cycles are shown in Figure 5 and in Table 3.

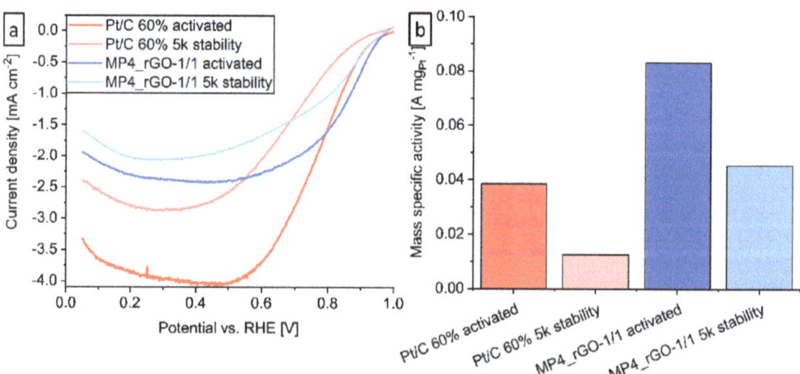

Figure 5. Electrochemical activity of the sample **MP4_rGO-1/1** and Pt/C 60% as reference: (**a**) obtained linear sweep voltammetry at 10 mV s^{-1} sweep rate after the activation and after 5k cycles; (**b**) calculated mass-specific activity at 0.9 V after activation and after 5k cycles.

Table 3. Mass-specific activity at 0.9 V of the sample **MP4_rGO-1/1** after activation and after 5k cycles.

Sample	MA [mA mg$_{Pt}^{-1}$]
Pt/C 60% activated	38
Pt/C 60% stability after 5k cycles	13
MP4_rGO-1/1 activated	83
MP4_rGO-1/1 stability after 5k cycles	45

The LSV of **MP4_rGO-1/1** after 5k cycles shows a decrease in MA, by roughly 45%, while Pt/C 60% loses more than 60% of MA. The activity loss of the sample can be explained with the aid of TEM images obtained after the stability test, shown in Figure 6. When compared with the freshly prepared NPs (Figure 3), the Pt-NP sample lost its small-size homogeneity, with also larger NPs, for which sizes of over 20 nm were clearly visible. Since this increase in size was not observed via PXRD and TEM for dried Pt-NPs, which were stored in air at room temperature for half a year and up to one year, we assume that this NP growth is linked to the ORR.

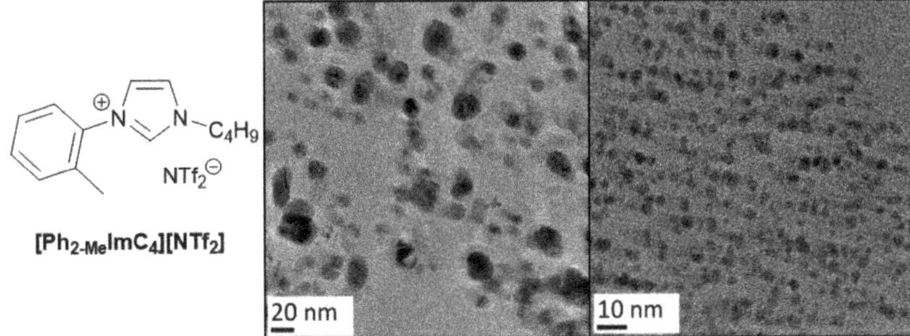

Figure 6. TEM images of the sample **MP4_rGO-1/1** obtained after 5k stability test. Images were measured with an FEI Titan 80–300 TEM.

We conclude that the NPs can be detached from the stabilizing rGO support during potential cycling and start to agglomerate, losing their electrochemical activity. Since small Pt-NPs remain on the GC surface and show electrochemical activity, the exact position of the NP on the rGO might have an influence on this process, with NPs attached to inner layers of the rGO being more stable. Although not visible, in images with small Pt-NPs, the resolution to depict the rGO can be difficult because stacked or strongly wrinkled rGO sheets might have been removed during ink preparation or electrocatalytic reactions. A complete detachment of Pt-NPs of the rGO would be unlikely because the activity loss would be much stronger with more NP agglomerates and no small NPs detectable.

3. Conclusions

Small Pt-nanoparticles (Pt-NPs) as dispersed in ionic liquids and in situ deposited on reduced graphite oxide (rGO) (Pt-NP@rGO) can be obtained during a fast and effective microwave-assisted thermal decomposition of (MeCp)PtMe$_3$ without and with rGO added and tunable aryl alkyl ionic liquids (TAAILs), as solvent and stabilizer. The particle size of the obtained Pt-NPs can vary slightly, depending on the TAAIL, and lies between 2 ± 1 nm and 6 ± 2 nm. Independent of the particle size, the Pt-NPs build dense aggregates. When comparing the TAAILs as reaction media with each other, the length of the alkyl chain or the aryl group only has an insignificant effect on the obtained particle size and aggregation.

For samples containing rGO, TEM images show that Pt-NPs build a layer on top of the rGO surface and maintain a homogenous size. In addition, no particle aggregation is observable, which demonstrates that rGO effectively anchors the obtained NPs and prevents their agglomeration. Pt-NPs and Pt-NP@rGO samples show good stability over at least 12 months, with no increase in particle size and no further agglomeration observable.

Pt-NP@rGO samples on glassy carbon electrodes for the oxygen reduction reaction (ORR) exhibit an up to a twofold increase in electrochemical surface area (ECSA) and mass-specific activity (MA), compared with Pt/C 60%. **MP5_rGO-2/1** achieved the best performance of the analyzed Pt-NP@rGO samples, which all appear promising for ORR and are worthy of further investigation. Since Pt-NP sizes between 2 nm and 3 nm are known to be the most active size range for ORR, and TAAILs make this size range readily available, their use as reaction media for Pt-NP preparation asks for further investigations. An organometallic Pt precursor and TAAILs in the presence of carbon support, in combination with microwave-induced heating, offer fast access to Pt-based electrocatalysts on the nanoscale.

4. Materials and Methods

4.1. Chemicals and Equipment

The following chemicals were received from commercial sources: Pt/C (60 wt% Pt on Vulcan XC-72R, Sigma-Aldrich), Nafion 1100 W (Sigma-Aldrich), n-butyllithium (1.6 mol L^{-1} in hexane, Acros organics), methyllithium (1.6 mol L^{-1} in diethyl ether, Sigma-Aldrich), 1,2-dibromoethane (>98%, Fluka), potassium hexachloridoplatinate (IV) (97%, BLDpharm), potassium iodide (USP, BP, Ph. Eur. pure, pharma grade, PanReac Applichem), methylcyclopentadienyl dimer (95%, Acros Organics), and perchloric acid (70%, ACS reagent, Sigma-Aldrich). All chemicals were used as received without further purification.

Thermally reduced graphite oxide (rGO) was prepared in a two-step oxidation–thermal reduction process using natural graphite (type KFL 99.5 from AMG Mining AG, former Kropfmühl AG, Passau, Germany). The graphite oxidation procedure of Hummers and Offeman [28] was followed. rGO was obtained at a reduction temperature of 400 °C from graphite oxide. For further information, also see our previous studies [21,25].

Tuneable aryl alkyl ionic liquids (TAAILs) were obtained in a two-step synthesis at the group of Prof. Thomas Strassner, Technische Universität, Dresden. The first step was the alkylation of the aryl imidazoles by bromoalkanes to build the IL cations. As the second step, the anion (bromide) was exchanged with LiNTf$_2$. For detailed information, see Supporting Information in Refs. [15,30].

(η^5-methylcyclopentadienyl)trimethylplatinum(IV) ((MeCp)PtMe$_3$) was synthesized and characterized after a method described by Xue et al. [13,33].

Transmission electron microscopy (TEM) measurements were performed with a JEOL-2100 Plus, a Zeiss LEO912, and an FEI Titan 80–300 TEM at 200 kV, 120 kV, and 300 kV accelerating voltage, respectively. The samples were prepared using 200 μm carbon-coated copper grids. Briefly, 0.05 mL of the NP/IL dispersion was dissolved in a 0.5 mL acetonitrile (ACN), and one drop of the diluted dispersion was placed on the grid. After 30 min, the grid was washed with 3 mL ACN and air dried. The images were analyzed by Gatan Microscopy Suite version 3.3, and the particle size distribution was determined from at least 200 individual particles at different positions on the TEM grid, with the same magnifications.

Powder X-ray diffractograms (PXRDs) were measured at ambient temperature on a Bruker D2-Phaser using a flat sample holder and Cu-Kα radiation (λ = 1.54182 Å, 35 kV). Diffrac.Eva V4.2 was used to evaluate PXRD data. Particle sizes were calculated with the Scherrer equation (Equation (1)) as follows:

$$L = K \times \lambda / (\Delta(2\theta) \times \cos \theta_0) \tag{1}$$

where L is the average crystallite size, K is the dimensionless shape factor, λ is the wavelength, $\Delta(2\theta)$ is the full width at half maximum (FWHM) in radians, and θ is the Bragg angle.

A CEM-Discover SP microwave setup, with a power range of 0–300 W ($\pm$30 W) was used for all microwave reactions. Thermogravimetric analysis (TGA) was conducted with a Netzsch TG 209 F3 Tarsus device, equipped with an Al crucible applying a heating rate of 5 °C/min under a synthetic air atmosphere.

4.2. Synthesis of Pt-NPs in IL

As a general approach to obtain Pt-NPs, (MeCp)PtMe$_3$ was put in a 10 mL microwave vessel and dispersed in the corresponding IL. The necessary amount of Pt precursor compared with IL was determined by the wt% of obtained Pt-NPs in IL with 100% conversion, which was set between 1 wt% and 2 wt%. The general batch size was further customized with roughly 300 mg, 500 mg, and 1.00 g IL. The dispersion was stirred for at least 6 h and then heated in the microwave (200 °C, 40 W, 10 min holding time). Afterward, 3 mL acetonitrile was added to the black dispersion, mixed, and centrifuged. The clear liquid phase was removed. This washing process was repeated until a colorless, clean solution could be removed (mostly four times). The residue was dried in a high vacuum, to obtain pure Pt-NPs. A list of all obtained Pt-NPs can be found in Table S1 in the Supplementary Information.

For the synthesis of Pt-NPs in IL with rGO, 0.5 wt% to 1 wt% rGO was additionally added into the microwave vessel, stirred overnight, and sonicated for 3 h. The following steps were carried out as mentioned before.

4.3. Electrochemical Measurements

For all measurements, a similar setup and procedure as in Beermann et al. were used [38]. A conventional three-electrode cell, with a Pt gauze as a counter electrode (Pt furled Pt 5 × 5 cm^2 mesh), a reference electrode (reversible hydrogen electrode = RHE), and a glassy carbon-working electrode (5 mm diameter), was used. The working electrode was always lowered into the electrolyte under potential control at 0.05 V$_{RHE}$. For all measurements, a 0.1 M HClO$_4$ electrolyte solution, diluted from 70% concentrated HClO$_4$ with Milli-Q water, was used. For monitoring, BioLogics potentiostats SP-150 or SP-200 were used.

Fresh inks were prepared from the NPs as the electrochemically active material. For the Pt-NP@rGO inks, NPs containing 1 mg Pt were mixed with 0.5 mL isopropanol, 2.5 mL water, and 10 µL nafion 5% and sonicated for at least 30 min. Since the samples already contained rGO as carbon support, the catalyst ink did not contain additional carbon. Next, 10 µL of the ink was deposited onto the working electrode, to achieve a loading of 10 µg$_{Pt}$ cm^{-2}, and dried for 10 min.

After the electrolyte was purged with N$_2$ for 20 min, the electrochemical activation was performed via potential cycling between 0.050 and 0.925 V$_{RHE}$, with a scan rate of 100 mV s^{-1} for 50 cycles under a N$_2$-protected atmosphere. The H-adsorption-based electrochemically active surface area (ECSA) was determined, with the last cycle recorded as 20 mV s^{-1}, before the activity measurements were selected. The cyclic voltammetry (CV) was carried out by cycling between 0.05 and 1.0 V$_{RHE}$, with a scan rate of 20 mV s^{-1}, under a N$_2$ atmosphere. The charge values (Q$_H$) were calculated by integrating the respective CV between 0.05 V and 0.4 V. The measured Q$_H$ values were normalized with respect to the theoretical value of Q$_H$^{theo} = 210 µC cm^{-2}, which is assuming a one-electron transfer between one H atom and one Pt atom. To determine the catalytic activity of the catalysts, linear sweep voltammetry (LSV) was measured in an oxygen saturated electrolyte (after 15 min of bubbling) in a potential range between 0.05 and 1.0 V$_{RHE}$, with a scan rate of 10 mV s^{-1} and a rotation speed of 1600 rpm. These measurements were repeated three times. The kinetic currents were calculated using (Equation (2)) as follows:

$$I_k = (I_{lim} \times I)/(I_{lim} - I) \qquad (2)$$

where I is the measured current density, I_{lim} is determined in the diffusion-limited current area, and I_k is the calculated kinetic current density. All presented current densities are iR corrected, where the uncompensated ohmic resistance (R) was determined by potential electrochemical impedance spectroscopy (PEIS), at 0.4 V$_{RHE}$. Additional rotation speeds of 225, 400, 625, 900, and 1225 rpm were measured for the sample **MP4_rGO-1/1**.

Stability measurements were performed via potential cycling between 0.5 and 1.0 V_{RHE}, with a scan rate of 50 mV s^{-1}, in N_2-saturated 0.1 M HClO$_4$, with 0 rpm. In total, 5k cycles were carried out. Before and after each stability protocol, three CV cycles between 0.05 and 1.0 V_{RHE} at a scan rate of 100 mV s^{-1} were recorded, followed by oxygen bubbling (15 min) and LSV, in a potential range between 0.05 and 1.0 V_{RHE}, with a scan rate of 10 mV s^{-1} and a rotation speed of 1600 rpm.

Supplementary Materials: The following supporting information can be downloaded. S1. Pt-NP and Pt-NP@rGO character-ization; S2. Powder X-ray diffractograms (PXRDs); S3. Transmission electron microscopy (TEM) images of Pt-NP and Pt-NP@rGO samples; S4. TGA measurements of selected Pt@rGO samples; S5. Electrocatalytic investigations of Pt@rGO samples.

Author Contributions: Conceptualization, C.J. and D.W.; methodology, D.W.; validation, D.W. and M.S.; formal analysis, D.W.; investigation, D.W., S.L., W.J., M.S. and S.R.; resources, C.J. and T.S.; data curation, D.W.; writing—original draft preparation, D.W.; writing—review and editing, C.J., M.S. and T.S.; visualization, D.W.; supervision, C.J.; project administration, C.J.; funding acquisition, C.J. and T.S. All authors have read and agreed to the published version of the manuscript.

Funding: The research of C.J. was supported by a joint National Natural Science Foundation of China–Deutsche Forschungsgemeinschaft (NSFC-DFG) project (DFG JA466/39-1). T.S. is grateful for funding by the Deutsche Forschungsgemeinschaft (SPP 1708, STR 526/20-1/2).

Institutional Review Board Statement: Not applicable.

Informed Consent Statement: Not applicable.

Data Availability Statement: The data presented in this study are available on request from the corresponding author.

Acknowledgments: We thank Rolf Mülhaupt and his group for providing the rGO material. We are grateful to Harry Biller for helpful discussions and Dennis Woschko for the TGA measure-ments. Access to the infrastructure at the Ernst Ruska-Centre, Forschungszentrum Jülich is gratefully acknowledged.

Conflicts of Interest: The authors declare that they have no known competing financial interest or personal relationship that could have appeared to influence the work reported in this paper.

References

1. Li, M.; Zhao, Z.; Cheng, T.; Fortunelli, A.; Chen, C.-Y.; Yu, R.; Zhang, Q.; Gu, L.; Merinov, B.; Lin, Z.; et al. Ultrafine jagged platinum nanowires enable ultrahigh mass activity for the oxygen reduction reaction. *Science* **2016**, *354*, 1414–1419. [CrossRef] [PubMed]
2. Mahata, A.; Nair, A.S.; Pathak, B. Recent advancements in Pt-nanostructure-based electrocatalysts for the oxygen reduction reaction. *Catal. Sci. Technol.* **2019**, *9*, 4835–4863. [CrossRef]
3. Lai, J.; Guo, S. Design of Ultrathin Pt-Based Multimetallic Nanostructures for Efficient Oxygen Reduction Electrocatalysis. *Small* **2017**, *13*, 1702156. [CrossRef] [PubMed]
4. Calderón, J.C.; Ndzuzo, L.; Bladergroen, B.J.; Pasupathi, S. Catalytic activity of carbon supported-Pt-Pd nanoparticles toward the oxygen reduction reaction. *Mater. Today Proc.* **2018**, *5*, 10551–10560. [CrossRef]
5. Shao, M.; Chang, Q.; Dodelet, J.-P.; Chenitz, R. Recent Advances in Electrocatalysts for Oxygen Reduction Reaction. *Chem. Rev.* **2016**, *116*, 3594–3657. [CrossRef]
6. Bao, J.; Wang, J.; Zhou, Y.; Hu, Y.; Zhang, Z.; Li, T.; Xue, Y.; Guo, C.; Zhang, Y. Anchoring ultrafine PtNi nanoparticles on N-doped graphene for highly efficient hydrogen evolution reaction. *Catal. Sci. Technol.* **2019**, *9*, 4961–4969. [CrossRef]
7. Chen, H.; Wang, G.; Gao, T.; Chen, Y.; Liao, H.; Guo, X.; Li, H.; Liu, R.; Dou, M.; Nan, S.; et al. Effect of Atomic Ordering Transformation of PtNi Nanoparticles on Alkaline Hydrogen Evolution: Unexpected Superior Activity of the Disordered Phase. *J. Phys. Chem. C* **2020**, *124*, 5036–5045. [CrossRef]
8. Zhang, H.; Bo, X.; Guo, L. Electrochemical preparation of Pt nanoparticles supported on porous graphene with ionic liquids: Electrocatalyst for both methanol oxidation and H$_2$O$_2$ reduction. *Electrochim. Acta* **2016**, *201*, 117–124. [CrossRef]
9. Cao, Y.; Yang, Y.; Shan, Y.; Huang, Z. One-Pot and Facile Fabrication of Hierarchical Branched Pt-Cu Nanoparticles as Excellent Electrocatalysts for Direct Methanol Fuel Cells. *ACS Appl. Mater. Interfaces* **2016**, *8*, 5998–6003. [CrossRef]
10. Ouyang, Y.; Cao, H.; Wu, H.; Wu, D.; Wang, F.; Fan, X.; Yuan, W.; He, M.; Zhang, L.Y.; Li, C.M. Tuning Pt-skinned PtAg nanotubes in nanoscales to efficiently modify electronic structure for boosting performance of methanol electrooxidation. *Appl. Catal. B Environ.* **2020**, *265*, 118606. [CrossRef]

11. Xie, C.; Niu, Z.; Kim, D.; Li, M.; Yang, P. Surface and Interface Control in Nanoparticle Catalysis. *Chem. Rev.* **2020**, *120*, 1184–1249. [CrossRef] [PubMed]
12. Rück, M.; Bandarenka, A.; Calle-Vallejo, F.; Gagliardi, A. Oxygen Reduction Reaction: Rapid Prediction of Mass Activity of Nanostructured Platinum Electrocatalysts. *J. Phys. Chem. Lett.* **2018**, *9*, 4463–4468. [CrossRef] [PubMed]
13. Marquardt, D.; Barthel, J.; Braun, M.; Ganter, C.; Janiak, C. Weakly-coordinated stable platinum nanocrystals. *CrystEngComm* **2012**, *14*, 7607–7615. [CrossRef]
14. Wegner, S.; Janiak, C. Metal Nanoparticles in Ionic Liquids. *Top. Curr. Chem.* **2017**, *375*, 65. [CrossRef] [PubMed]
15. Schmolke, L.; Lerch, S.; Bülow, M.; Siebels, M.; Schmitz, A.; Thomas, J.; Dehm, G.; Held, C.; Strassner, T.; Janiak, C. Aggregation control of Ru and Ir nanoparticles by tunable aryl alkyl imidazolium ionic liquids. *Nanoscale* **2019**, *11*, 4073–4082. [CrossRef]
16. Ahrens, S.; Peritz, A.; Strassner, T. Tunable aryl alkyl ionic liquids (TAAILs): The next generation of ionic liquids. *Angew. Chem. Int. Ed.* **2009**, *48*, 7908–7910. [CrossRef]
17. Pârvulescu, V.I.; Hardacre, C. Catalysis in ionic liquids. *Chem. Rev.* **2007**, *107*, 2615–2665. [CrossRef]
18. Migowski, P.; Machado, G.; Texeira, S.R.; Alves, M.C.M.; Morais, J.; Traverse, A.; Dupont, J. Synthesis and characterization of nickel nanoparticles dispersed in imidazolium ionic liquids. *Phys. Chem. Chem. Phys.* **2007**, *9*, 4814–4821. [CrossRef]
19. He, F.-G.; Yin, J.-Y.; Sharma, G.; Kumar, A.; Stadler, F.J.; Du, B. Facile Fabrication of Hierarchical rGO/PANI@PtNi Nano-composite via Microwave-Assisted Treatment for Non-Enzymatic Detection of Hydrogen Peroxide. *Nanomaterials* **2019**, *9*, 1109. [CrossRef]
20. Marquardt, D.; Beckert, F.; Pennetreau, F.; Tölle, F.; Mülhaupt, R.; Riant, O.; Hermans, S.; Barthel, J.; Janiak, C. Hybrid materials of platinum nanoparticles and thiol-functionalized graphene derivatives. *Carbon* **2014**, *66*, 285–294. [CrossRef]
21. Schmitz, A.; Schütte, K.; Ilievski, V.; Barthel, J.; Burk, L.; Mülhaupt, R.; Yue, J.; Smarsly, B.; Janiak, C. Synthesis of metal-fluoride nanoparticles supported on thermally reduced graphite oxide. *Beilstein J. Nanotechnol.* **2017**, *8*, 2474–2483. [CrossRef]
22. Kim, I.G.; Nah, I.W.; Oh, I.-H.; Park, S. Crumpled rGO-supported Pt-Ir bifunctional catalyst prepared by spray pyrolysis for unitized regenerative fuel cells. *J. Power Sour.* **2017**, *364*, 215–225. [CrossRef]
23. Lee, G.; Shim, J.H.; Kang, H.; Nam, K.M.; Song, H.; Park, J.T. Monodisperse Pt and PtRu/C_{60} hybrid nanoparticles for fuel cell anode catalysts. *Chem. Commun.* **2009**, 5036–5038. [CrossRef]
24. Hao, Y.; Wang, X.; Zheng, Y.; Shen, J.; Yuan, J.; Wang, A.-J.; Niu, L.; Huang, S. Size-controllable synthesis of ultrafine PtNi nanoparticles uniformly deposited on reduced graphene oxide as advanced anode catalysts for methanol oxidation. *Int. J. Hydrogen Energy* **2016**, *41*, 9303–9311. [CrossRef]
25. Marquardt, D.; Vollmer, C.; Thomann, R.; Steurer, P.; Mülhaupt, R.; Redel, E.; Janiak, C. The use of microwave irradiation for the easy synthesis of graphene-supported transition metal nanoparticles in ionic liquids. *Carbon* **2011**, *49*, 1326–1332. [CrossRef]
26. Zhang, B.; Ning, W.; Zhang, J.; Qiao, X.; Zhang, J.; He, J.; Liu, C.-Y. Stable dispersions of reduced graphene oxide in ionic liquids. *J. Mater. Chem.* **2010**, *20*, 5401–5403. [CrossRef]
27. Öztürk, S.; Xiao, Y.-X.; Dietrich, D.; Giesen, B.; Barthel, J.; Ying, J.; Yang, X.-Y.; Janiak, C. Nickel nanoparticles supported on a covalent triazine framework as electrocatalyst for oxygen evolution reaction and oxygen reduction reactions. *Beilstein J. Nanotechnol.* **2020**, *11*, 770–781. [CrossRef]
28. Hummers, W.S., Jr.; Offeman, R.E. Preparation of Graphitic Oxide. *J. Am. Chem. Soc.* **1958**, *80*, 1339. [CrossRef]
29. Lerch, S.; Strassner, T. Expanding the Electrochemical Window: New Tunable Aryl Alkyl Ionic Liquids (TAAILs) with Dicyanamide Anions. *Chem.-Eur. J.* **2019**, *25*, 16251–16256. [CrossRef]
30. Lerch, S.; Strassner, T. Synthesis and Physical Properties of Tunable Aryl Alkyl Ionic Liquids (TAAILs). *Chem.-Eur. J.* **2021**, *27*, 15554–15557. [CrossRef]
31. Thurier, C.; Doppelt, P. Platinum OMCVD processes and precursor chemistry. *Coord. Chem. Rev.* **2008**, *252*, 155–169. [CrossRef]
32. Lubers, A.M.; Muhich, C.L.; Anderson, K.M.; Weimer, A.W. Mechanistic studies for depositing highly dispersed Pt na-noparticles on carbon by use of trimethyl(methylcyclopentadienyl)platinum(IV) reactions with O_2 and H_2. *J. Nanoparticle Res.* **2015**, *17*, 85. [CrossRef]
33. Xue, Z.; Strouse, M.J.; Shuh, D.K.; Knobler, C.B.; Kaesz, H.D.; Hicks, R.F.; Williams, R.S. Characterization of (methylcyclopentadienyl)trimethylplatinum and low-temperature organometallic chemical vapor deposition of platinum metal. *J. Am. Chem. Soc.* **1989**, *111*, 8779–8784. [CrossRef]
34. Badam, R.; Vedarajan, R.; Okaya, K.; Matsutani, K.; Matsumi, N.J. Ionic Liquid Mediated Decoration of Pt Nanoparticles on Graphene and Its Electrocatalytic Characteristics. *J. Electrochem. Soc.* **2021**, *168*, 36515. [CrossRef]
35. Daş, E.; Öztürk, A.; Bayrakçeken Yurtcan, A. Electrocatalytical Application of Platinum Nanoparticles Supported on Reduced Graphene Oxide in PEM Fuel Cell: Effect of Reducing Agents of Dimethlyformamide or Hydrazine Hydrate on the Properties. *Electroanalysis* **2021**, *33*, 1721–1735. [CrossRef]
36. Teran-Salgado, E.; Bahena-Uribe, D.; Márquez-Aguilar, P.A.; Reyes-Rodriguez, J.L.; Cruz-Silva, R.; Solorza-Feria, O. Plat-inum nanoparticles supported on electrochemically oxidized and exfoliated graphite for the oxygen reduction reaction. *Electrochim. Acta* **2019**, *298*, 172–185. [CrossRef]

37. Holewinski, A.; Linic, S. Elementary Mechanisms in Electrocatalysis: Revisiting the ORR Tafel Slope. *J. Electrochem. Soc.* **2012**, *159*, H864–H870. [CrossRef]
38. Beermann, V.; Gocyla, M.; Kühl, S.; Padgett, E.; Schmies, H.; Goerlin, M.; Erini, N.; Shviro, M.; Heggen, M.; Dunin-Borkowski, R.E.; et al. Tuning the Electrocatalytic Oxygen Reduction Reaction Activity and Sta-bility of Shape-Controlled Pt-Ni Nanoparticles by Thermal Annealing—Elucidating the Surface Atomic Structural and Compositional Changes. *J. Am. Chem. Soc.* **2017**, *139*, 16536–16547. [CrossRef]

Article

PtNi Alloy Coated in Porous Nitrogen-Doped Carbon as Highly Efficient Catalysts for Hydrogen Evolution Reactions

Xuyan Song [1,2], Yunlu He [1], Bo Wang [1], Sanwen Peng [2], Lin Tong [1], Qiang Liu [1], Jun Yu [3,*] and Haolin Tang [3,*]

[1]   China Tobacco Hubei Industrial Co., Ltd., Wuhan 430051, China; songxy@hbtobacco.cn (X.S.); heyunlu@hbtobacco.cn (Y.H.); wangbo@hbtobacco.cn (B.W.); tonglin@hbtobacco.cn (L.T.); liuqiang@market.hbtobacco.cn (Q.L.)
[2]   China Tobacco Hubei Industrial Cigarette Materials Co., Ltd., Wuhan 430051, China; pengsw@market.hbtobacco.cn
[3]   State Key Laboratory of Advanced Technology for Materials Synthesis and Processing, Wuhan University of Technology, Wuhan 430070, China
*   Correspondence: yujun@whut.edu.cn (J.Y.); thln@whut.edu.cn (H.T.)

Abstract: The development of low platinum loading hydrogen evolution reaction (HER) catalysts with high activity and stability is of great significance to the practical application of hydrogen energy. This paper reports a simple method to synthesize a highly efficient HER catalyst through coating a highly dispersed PtNi alloy on porous nitrogen-doped carbon (MNC) derived from the zeolite imidazolate skeleton. The catalyst is characterized and analyzed by physical characterization methods, such as XRD, SEM, TEM, BET, XPS, and LSV, EIS, it, v-t, etc. The optimized sample exhibits an overpotential of only 26 mV at a current density of 10 mA cm^{-2}, outperforming commercial 20 wt% Pt/C (33 mV). The synthesized catalyst shows a relatively fast HER kinetics as evidenced by the small Tafel slope of 21.5 mV dec^{-1} due to the small charge transfer resistance, the alloying effect between Pt and Ni, and the interaction between PtNi alloy and carrier.

Keywords: hydrogen evolution reaction; porous carbon; PtNi alloy

Citation: Song, X.; He, Y.; Wang, B.; Peng, S.; Tong, L.; Liu, Q.; Yu, J.; Tang, H. PtNi Alloy Coated in Porous Nitrogen-Doped Carbon as Highly Efficient Catalysts for Hydrogen Evolution Reactions. *Molecules* **2022**, *27*, 199. https://doi.org/10.3390/molecules27020499

Academic Editors: Jingqi Guan and Yin Wang

Received: 23 December 2021
Accepted: 10 January 2022
Published: 14 January 2022

Publisher's Note: MDPI stays neutral with regard to jurisdictional claims in published maps and institutional affiliations.

1. Introduction

Hydrogen has the advantages of having high energy density and being clean and pollution-free, making it an ideal energy source to replace traditional fossil fuels to solve environmental and energy issues [1–3]. The development of an environmentally sound and sustainable hydrogen production method is the basis for the application of hydrogen energy. Hydrogen production by electrolysis of water has received extensive attention due to its clean and environmental advantages. The hydrogen evolution reaction is the key reaction to produce hydrogen through water electrolysis. To date, platinum (Pt) and Pt-based catalysts are still the ideal catalysts for the hydrogen evolution reaction because of their high catalytic activity and long-term stability. Unfortunately, the high cost and limited reserves hinder its large-scale promotion and commercial use.

Improving the catalytic activity and utilization rate of precious platinum to reduce the usage of platinum has become the key measure to break this deadlock. Supported platinum-based alloys with a high specific surface area is a research focus in reducing the content of platinum in the catalyst and improving the catalytic activity of the catalyst [4,5]. Benefiting from the synergistic effect induced by the interaction of Pt and transition metals, the combination of Pt with transition metals (Fe, Co, Ni, etc.) to form a composite alloy is an effective way to optimize the utilization of Pt and to improve its electrocatalytic activity [6,7]. As an example, Huang et al. [8] synthesized PtNi nanodentrites (PtNi NDs) by a simple solvothermal method. The transition metal Ni can adjust the inherent electronic structure of Pt to improve the catalytic activity. In 0.5 M H$_2$SO$_4$, the optimal sample PtNi NDs requires only 22 mV overpotential under current density of 10 mA cm^{-2}, better than commercial 20 wt% Pt/C (30 mV) under the same condition. The resulting Tafel slope of 52 mV dec $^{-1}$ is much lower than that of 20 wt% Pt/C (66 mV dec^{-1}).

Metal-organic framework (MOFs) materials have attracted great attention in the field of catalysis due to their high specific surface area and easy regulation of pore size and coordination center of metal atoms [9]. Meanwhile, derivatives of MOFs materials as the carrier materials of Pt-based catalysts can greatly improve the utilization of the precious metal Pt, thereby achieving the purpose of reducing the amount of the precious metal. As a branch of MOFs material, ZIFs (zeolite imidazolium ester skeleton structure materials) is widely used in the field of electrolytic water catalysis [10,11]. For example, Qin et al. [12] synthesized the PtCo bimetallic catalyst with ZIF-67 as a porous carbon source through the simple impregnation method. The Pt content of the optimal sample CPt@ZIF-67-900-6 is only 5 wt%. When the current density reaches 10 mA cm^{-2}, the overpotential of CPt@ZIF-67-900-6(50 mV) is 5 mV lower than that of commercial 20 wt% Pt/C and the Tafel slope of CPt@ZIF-67-900-6 (27.1 mV dec^{-1}) is much lower than that of the 20 wt%Pt/C catalyst (35.5 mV dec^{-1}). The density functional theory calculation proves that ZIFs derivatives can improve the utilization rate of Pt.

Inspired by the above-mentioned analysis, porous carbon (MNC) support is prepared in this paper by the carbonization of Co-doped ZIF-8. The thus-obtained porous carbon not only has the advantage of a large specific surface area, but also contains a large number of Co-N$_x$ active sites, which can enhance the catalytic activity [13,14]. Then, the electrocatalyst PtNi-MNC-x-y (x:y is the mass ratio of Co and Zn) is prepared by supporting platinum-nickel alloy on the porous carbon through a simple impregnation method. The effect of synthetic conditions on the microstructure and performance of materials is discussed in detail. The results revealed that the sample derived from a carbonization temperature of 900 °C and Co:Zn of 1:9 displayed the best performance due to the small charge transfer resistance, the alloying effect between Pt and Ni, and the interaction between PtNi alloy and the carrier.

The experimental details and electrochemical measurements are described in the Supplementary Material.

2. Results and Discussion

2.1. Material Characterization

X-ray diffraction technology was first applied to analyze the crystal structure of the sample. Figure 1 shows XRD patterns of PtNi/MNC-1-6 and Pt/MNC-1-6. Although the Zn element was added during the materials preparation process, it can be almost completely removed as confirmed by the XRD patterns and the ICP-AES results (Table 1). In Figure 1, it can be observed that there are three diffraction peaks at 39.8°, 46.2° and 67.7°, corresponding to crystal planes of Pt (111), Pt (200), and Pt (220), respectively (ICDD 04-0802). Moreover, the observed slightly positive shift of Pt diffraction peaks (as shown by the arrow in the Figure 1) for PtNi/MNC-1-6 compared to the sample of Pt/MNC-1-6 suggested the change of lattice spacing of Pt after the formation of the PtNi alloy, confirming the formation of the PtNi alloy. In addition, three diffraction peaks of Co element can also be observed at 44.2°, 51.5° and 75.9° corresponding to Co (111), Co (200) and Co (220) crystal planes (ICDD 15-0806), respectively. A certain amount of Co–N$_x$ can be formed during the pyrolysis process, which has been proved to be the active site to the electrolytic reaction [15]. The position of Co diffraction peaks has no obvious shift from the peak of the standard metal Co, indicating that the Co element does not form an alloy with Pt or Ni metal. Moreover, there is no obvious diffraction peak of nickel, which is speculated to be caused by the low content of nickel element (2.9 wt % as shown in Table 1). Additionally, Co (ICDD 15-0806) and Ni (ICDD 04-0850) have similar peak positions, which made it difficult to show the diffraction peak of nickel clearly.

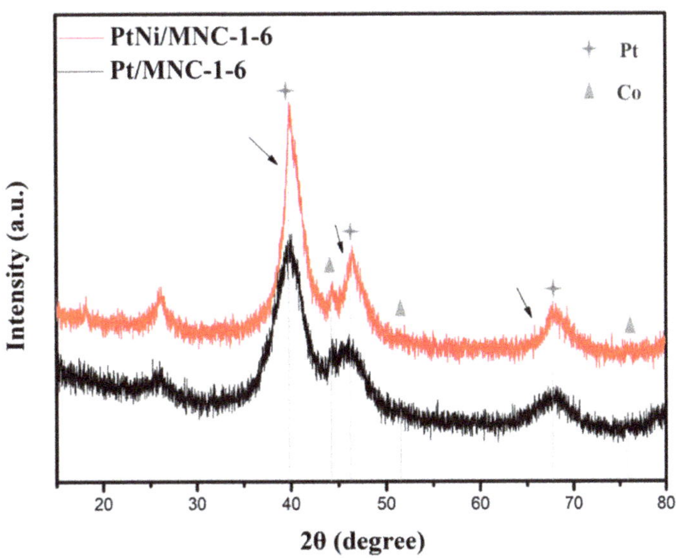

Figure 1. XRD patterns of PtNi/MNC-1-6 and Pt/MNC-1-6.

Table 1. Content test of each element in the PtNi/MNC-1-6 sample.

Sample	Measured Element	Mass Fraction (wt%)
PtNi/MNC-1-6	Pt	8.1
	Ni	2.9
	Co	12
	Zn	0.022

Scanning electron microscopy (SEM) and transmission electron microscopy (TEM) were applied to analyze the morphology of the catalyst, as shown in Figure 2a,b. It can be observed that the synthesized Co@ZIFs has a relatively uniform dodecahedral shape with a diameter of about 250 nm, indicating that most of the Co element was located on the surface of ZIF-8. As shown in Figure 2g, the outer surface of Co@ZIFs-1-6 is the Co element, and the inner part is the Zn element, which is basically consistent with the expectation. Meanwhile, it can be observed that the morphology of Co@ZIFs-1-3 does not achieve the expected results. It can be clearly seen in Figure S1a that its size is extremely uneven, which may be due to the excessive addition of Co elements, part of which form an independent ZIF-67 monomer, resulting in different sizes. After carbonization, the dodecahedral morphology of Co@ZIFs-1-6 is retained, but the surface becomes rough and the size is slightly reduced due to carbonization of organic bonds and atomic migration [4]. It is noteworthy that nanotubular structures in MNC-1-6 (Figure 2c and Supplementary Material S1c) were observed, possibly caused by the catalytic action of the metal Co on the surface of Co@ZIFs-1-6 during the carbonization process to form carbon nanotubes (CNTs) [5,16]. The generated CNTs are expected to be beneficial for the improvement of the specific surface area, catalytic activity, and stability of catalyst materials [16,17]. In addition, the numbers of the generated CNTs increased with the increases in the Co contents in the initial precursors. Figure 2d shows the morphology of PtNi/MNC-1-6. It can be seen that there are many deposited particles on the carbon carrier. In order to further explore the specific conditions of these sediments, electron transmission electron microscope (TEM) images were recorded, as shown in Figure 2e,f. The lattice spacing of 0.34 nm and 0.20 nm in Figure 2e corresponds to the C (002) plane and Co (111) plane and Co nanoparticles are coated with 8–10 layers of graphite-type carbon, which enhances the conductivity of the catalyst and the corrosion resistance of the metal particles coated by the carbon layer, thus improving the stability of the catalyst [18]. The lattice spacing of the particles shown in Figure 2f is 0.181 nm between the crystal plane of metal Pt (200) (0.196 nm) and metal Ni (200) (0.176 nm), and the lattice spacing of 0.212 nm is between the crystal plane of the corresponding metal Pt (111) (0.225 nm) and the metal Ni (111) (0.203 nm), confirming the formation of the PtNi alloys [19].

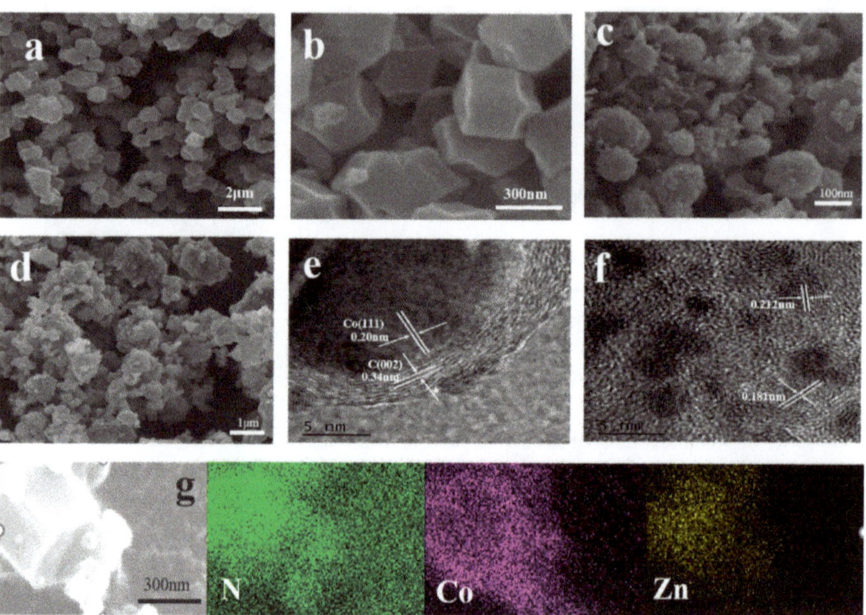

Figure 2. (**a**) SEM image of Co@ZIFs-1-6; (**b**) corresponding enlarged SEM image; (**c**) SEM image of MNC-1-6; (**d**) SEM image of PtNi/MNC-1-6; (**e**,**f**) TEM image of PtNi/MNC-1-6; (**g**) SEM image of Co@ZIFs-1-6 and its EDS image.

N_2 adsorption and desorption isotherms were performed to determine the specific surface area and pore size distribution of samples. Figure 3 shows the N_2 adsorption and desorption isotherms of MNC-1-6-800 °C, MNC-1-6-900 °C, and MNC-1-6-1000 °C and the corresponding pore size distribution curve. It is apparent that all the three recorded N_2 adsorption and desorption isotherms exhibited H4 hysteresis loops, indicating that all the three carriers derived from different carbonization temperatures possess mesoporous structures, and the calculated pore sizes are mainly distributed at 3–5 nm. The derived specific surface area of MNC-1-6-800 °C, MNC-1-6-900 °C, and MNC-1-6-1000 °C are 251 m^2/g, 343 m^2/g and 119 m^2/g, respectively. It is evident that the sample obtained from carbonization temperature of 900 °C exhibited the largest specific surface area among the three tested samples. This could suggest that the incomplete carbonization induced a less rough surface at 800 °C and the agglomeration of the derived carbon materials at 1000 °C are responsible for the relatively low specific surface area [20]. Evidently, the large specific surface area is beneficial to generation of active sites after the deposition of catalysts, which in turn can improve the utilization rate of the precious metal and the catalytic activity of the catalyst. Therefore, the carbonization temperature of 900 °C can be realized as the optimal condition for the synthesis of the catalyst support.

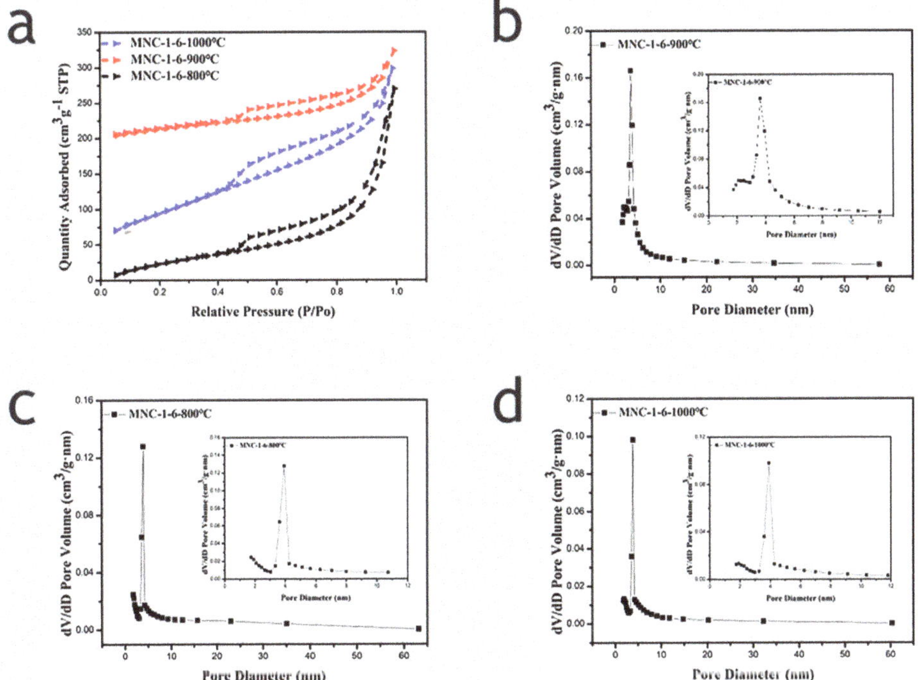

Figure 3. (**a**) N$_2$ adsorption and desorption isotherms of MNC-1-6-800 °C, MNC-1-6-900 °C and MNC-1-6-1000 °C; (**b–d**) the corresponding pore size distribution curve (Inset: Enlarged view of the corresponding small size pore size distribution).

X-ray photoelectron spectroscopy was carried out to elucidate the composition and chemical states of the according elements near the surface of the PtNi/MNC-1-6. The XPS survey was corrected by the C 1s peak fixed at 284.8 eV. As shown in Figure 4a, a high-resolution C 1s peak is deconvoluted into C–C (284.8 eV), C–N (286.0 eV) and C–C=O bonds (289.8 eV), indicating that the sample contains nitrogen-doped carbon and graphite carbon [21].The corresponding N 1s peak can be deconvoluted into four peaks at 398.4 eV, 399.7 eV, 401.1 eV and 405.5 eV, corresponding to pyridine N, pyrrole N, graphite N and oxide N, respectively, among which pyrrole N and pyridine N are essential for the electrocatalytic activity of HER by interacting with H$^+$ [16,21]. As shown in Figure 4d, Co 2p peak is deconvoluted and integrated into three pairs of peaks, among which the characteristic peaks at the binding energy of 778.4 eV and 789.1 eV are assigned to Co^0, indicating the existence of metallic cobalt. The characteristic peaks at 781.7 eV and 797.7 eV correspond to Co 2p$_{3/2}$ and Co 2p$_{1/2}$, respectively, suggesting the presence of Co^{2+} in the sample, possibly due to the surface oxidation of the catalyst during the test [22]. The other peaks correspond to the satellite peaks of Co. Similarly, Figure 4e shows the deconcolution of the Ni 2p peak, the characteristic peaks of Ni^{2+} at 856.2 eV for Ni 2p$_{3/2}$ and 873.8 eV for Ni 2P$_{1/2}$ and the characteristic peaks for Ni^0 at 852.3 eV and 871.2 eV. It is also found that the content of Ni^{2+} is slightly higher than Ni^0, which is basically consistent with the results reported in the previous article [23,24]. In Figure 4c, the Pt 4f in the PtNi/MNC-1-6 sample contains two pairs of characteristic peaks, corresponding to the two valence states of the Pt element. The characteristic peaks at 71.2 eV and 74.6 eV are Pt 4f$_{7/2}$ and Pt 4f$_{5/2}$ belonging to Pt^0, while the peaks at 72.1 eV and 75.7 eV are attributed to Pt^{2+} caused by surface oxidation [25,26].

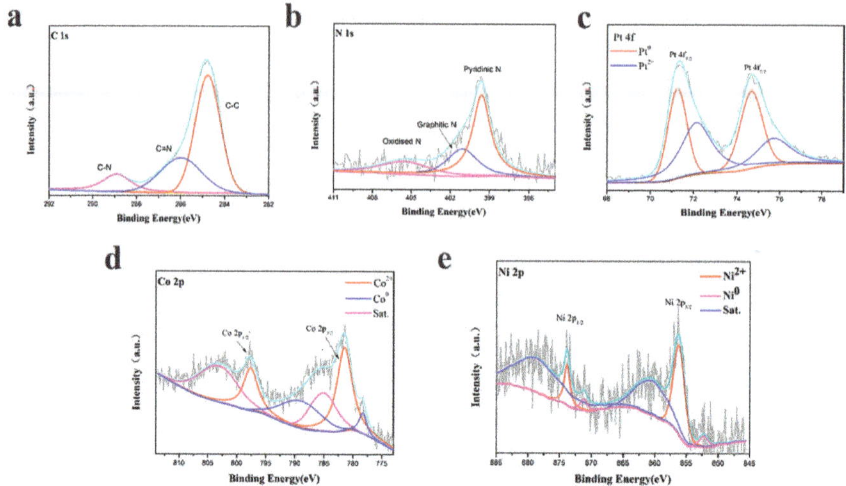

Figure 4. XPS spectra of PtNi/MNC-1-6: (**a**) C 1s; (**b**) N 1s; (**c**) Pt 4f; (**d**) Co 2p; (**e**) Ni 2p.

2.2. Electrochemical Activity

The influence of the carbonization temperature on the performance of the catalyst was investigated through the polarization curves of samples derived from different carbonization temperatures. As shown in Figure 5a,b, the required overpotentials for reaching a current density of 10 mA cm^{-2} are 33 mV, 26 mV and 36 mV for PtNi/MNC-1-6-800 °C, PtNi/MNC-1-6-900 °C and PtNi/MNC-1-6-1000 °C, respectively, suggesting that PtNi/MNC-1-6-900 °C exhibits the best performance among the tested samples, due to its high surface area, which enables the uniform distribution of the PtNi alloy on the surface of support. In addition, the relatively low carbonization temperature of 800 °C leads to the low content of graphitized carbon, which accordingly results in low conductivity and the catalytic performance [27]. Tafel curves derived from polarization curves were plotted in Figure 5c to explore the kinetics of the HER process for the three samples. It was observed that the Tafel slope of PtNi/MNC-1-6-900 °C is only 21.5 mV dec^{-1}, much lower than 30.7 mV dec^{-1} for PtNi/MNC-1-6-800 °C and 27.5 mV dec^{-1} for PtNi/MNC-1-6-1000 °C, indicating that the current density of sample PtNi/MNC-1-6-900 °C increased much faster with the increase in overpotential compared to the other two samples, implying the fast electrochemical reaction kinetics [23].

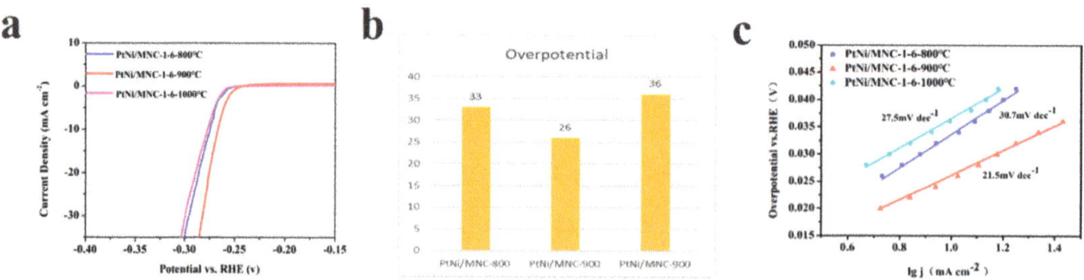

Figure 5. (**a**) HER polarization curve of the carrier-supported catalysts with different carbonization temperatures (polarization curves without i–R correction recorded in a 0.5 M H$_2$SO$_4$ solution at a sweep rate of 5 mV s^{-1}); (**b**) the corresponding overpotential; (**c**) the fitted Tafel curve of the corresponding polarization curve.

The influence of the introduced Ni to the Pt catalyst on the HER performance was also investigated. As shown in Figure 6a,b, the sample of PtNi/MNC-1-6 only needs the overpotential of 26 mV to reach a current density 10 mA cm^{-2}, superior to Pt/MNC-1-6 (35 mV). In addition, the advantage of the nickel-containing sample is also reflected by the Tafel curves. As can be seen from Figure 6c, the Tafel slope of sample PtNi/MNC-1-6 is 21.5 mV dec^{-1}, which is also lower than that of Pt/MNC-1-6 (23.3 mV dec^{-1}). It can also be clearly seen from the impedance spectra that the diameter of the impedance semicircle for sample PtNi/MNC-1-6 is much smaller than that for Pt/MNC-1-6, indicating that sample PtNi/MNC-1-6 has a much smaller charge transfer resistance (Rct) than Pt/MNC-1-6 does. The results demonstrate that the introduced transition metal Ni can adjust the intrinsic electronic structure of Pt and improve the catalytic activity [8,28]. In Table S1, we have listed summary of various PtNi alloys electrocatalysts for HER performance in 0.5 M H$_2$SO$_4$.

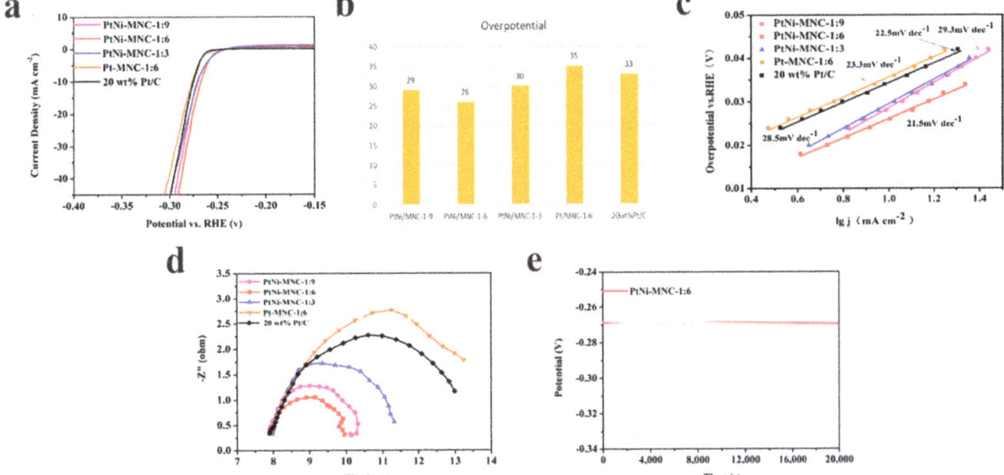

Figure 6. (**a**) Polarization curves of various catalysts (polarization curves without i–R correction recorded in a 0.5 M H$_2$SO$_4$ solution at a sweep rate of 5 mV s^{-1}); (**b**) corresponding overpotential image of catalysts; (**c**) the fitted Tafel curve of the corresponding polarization curve; (**d**) electrochemical impedance spectroscopy of various catalysts and 20 wt% Pt/C samples; (**e**) the chronopotential curve of PtNi/MNC-1-6.

The influence of the ratio of Co to Zn in the support on the performance of the catalyst was explored. As shown in Figure 6a,b, to reach a current density 10 mA cm^{-2}, the overpotential of sample PtNi/MNC-1-6 is 26 mV, about 4 mV and 3 mV lower than those of sample PtNi/MNC-1-3 (30 mV) and sample PtNi/MNC-1-9 (29 mV), respectively. In addition, the overpotentials of all the three samples are lower than that of commercial 20 wt% Pt/C (33 mV), possibly attributed to the Pt–Ni interaction and PtNi-alloy-support interaction. The best performance of sample PtNi/MNC-1-6 among all the tested samples is mainly attributed to the appropriate Co and Zn mass ratio induced porous morphology and the amount of generated CNTs as discussed in the above section.

The derived Tafel curve from the polarization curve is applied to elucidate the kinetics of the HER process. Under acidic conditions, hydrogen evolution reaction is generally analyzed by two mechanisms in three steps [29,30]. The three steps are the Volmer, Heyrovsky, and Tafel steps, and the corresponding Tafel slope is 120 mV dec^{-1}, 40 mV dec^{-1} and 30 mV dec^{-1}, respectively. The two mechanisms are the Volmer–Tafel mechanism and Volmer–Heyrovsky mechanism. The reaction of these three steps is as follows:

$$H_3O^+ + e^- \rightarrow H^* + H_2O \text{ (Volmer)} \tag{1}$$

$$H^* + H_3O^+ + e^- \rightarrow H_2 + H_2O \text{ (Heyrovsky)} \tag{2}$$

$$H^* + H^* \rightarrow H_2 \text{ (Tafel)} \tag{3}$$

H^*: H of adsorption state.

The Tafel slopes of PtNi/MNC-1-3, PtNi/MNC-1-6 and PtNi/MNC-1-9 are 28.5 mV dec^{-1}, 21.5 mV dec^{-1} and 29.3 mV dec^{-1}, respectively. It can be observed that the Tafel slope of sample PtNi/MNC-1-6 is lower than that of 20 wt% Pt/C (22.5 mV dec^{-1}), indicating that PtNi/MNC-1-6 exhibited faster electrochemical kinetics than commercial Pt/C. In addition, the HER process using the designed materials in this work follows the Volmer–Tafel mechanism.

Electrochemical impedance spectroscopy was applied to analyze the reaction kinetics of each catalyst. As shown in Figure 6d, the charge transfer resistance of PtNi/MNC-1-6 is much lower than that of PtNi/MNC-1-3, PtNi/MNC-1-9 and 20 wt% of Pt/C, indicating that PtNi/MNC-1-6 has lower charge transfer resistance and faster electrode kinetics [28]. The low charge transfer resistance is mainly due to appropriate carbonization temperatures and the strong interaction between PtNi and the support [31]. The chronopotentiometry technique was used to evaluate the stability of the catalyst. As shown in Figure 6e, the potential of the catalyst remained almost unchanged after continuous operation at the potential corresponding to 10 mA cm^{-2} for 20,000 s, demonstrating the great stability of PtNi/MNC-1-6.

3. Conclusions

In this work, the PtNi alloy was loaded on ZIF-derived carbon support as a catalyst for hydrogen evolution reaction and the influence of synthetic conditions and catalyst composition on the performance of the catalyst was studied. The catalyst was characterized and analyzed by physical characterization methods, such as XRD, SEM, TEM, BET, XPS, and LSV, EIS, it, v-t, etc. The optimal sample PtNi/MNC-1-6 (Pt content 8.1%) requires only 26 mV overpotential to reach a current density of 10 mA cm^{-2} with a small Tafel slope of 21.5 mV dec^{-1}. Moreover, the sample shows good stability. The excellent performance of the synthesized sample mainly benefits from the following points. (1) The proper Co and Zn ratio and carbonization temperature of the carrier provide good conductivity and a large specific surface area, which is conducive to the full dispersion of precious metals and improve the utilization rate of precious metals. (2) The synergistic effect of metal Pt and Ni improves the catalytic activity of the catalyst. (3) The interaction between the PtNi alloy and carbon support is beneficial to the rapid transfer of electrons.

Supplementary Materials: The following are available online: Experimental section; Figure S1 and Table S1.

Author Contributions: Conceptualization, X.S. and Y.H.; methodology, X.S., B.W. and Y.H.; software, X.S. and Y.H.; validation, X.S., B.W. and Y.H.; formal analysis, X.S., B.W. and S.P.; investigation, X.S., L.T. and S.P.; resources, X.S., Q.L., J.Y. and H.T.; writing original draft preparation, X.S.; All authors have read and agreed to the published version of the manuscript.

Funding: Fundamental Research Funds for the Central Universities, WUT: 2017III055, 2018III039GX, 2018IVA095.

Institutional Review Board Statement: Not applicable.

Informed Consent Statement: Not applicable.

Data Availability Statement: Not applicable.

Acknowledgments: This work was financed by the National Natural Science Foundation of China (Grant No. 22075223) and the Fundamental Research Funds for the Central Universities (WUT: 2017III055, 2018III039GX, 2018IVA095). We thank the Materials Analysis Center of Wuhan University of Technology for test supports.

Conflicts of Interest: There are no conflict of interest to declare.

Sample Availability: Samples are available from the authors.

References

1. Ishaq, H.; Dincer, I. Comparative assessment of renewable energy-based hydrogen production methods. *Renew. Sustain. Energy Rev.* **2021**, *135*, 110192. [CrossRef]
2. Dawood, F.; Anda, M.; Shafiullah, G.M. Hydrogen production for energy: An overview. *Int. J. Hydrogen Energy* **2020**, *45*, 3847–3869. [CrossRef]
3. Zhang, F.; Zhao, P.; Niu, M.; Maddy, J. The survey of key technologies in hydrogen energy storage. *Int. J. Hydrogen Energy* **2016**, *41*, 14535–14552. [CrossRef]
4. Torad, N.; Hu, M.; Ishihara, S.; Sukegawa, H.; Belik, A.; Imura, M.; Ariga, K.; Sakka, Y.; Yamauchi, Y. Direct Synthesis of MOF-Derived Nanoporous Carbon with Magnetic Co Nanoparticles toward Efficient Water Treatment. *Small* **2014**, *10*, 2096–2107. [CrossRef] [PubMed]
5. Xia, B.Y.; Yan, Y.; Li, N.; Wu, H.B.; Lou, X.W.; Wang, X. A metal-organic framework-derived bifunctional oxygen electrocatalyst. *Nat. Energy* **2016**, *1*, 15006. [CrossRef]
6. Yang, T.; Zhu, H.; Wan, M.; Dong, L.; Zhang, M.; Du, M. Highly efficient and durable PtCo alloy nanoparticles encapsulated in carbon nanofibers for electrochemical hydrogen generation. *Chem. Commun.* **2016**, *52*, 990–993. [CrossRef] [PubMed]
7. Lai, J.P.; Luque, R.; Xu, G.B. Recent Advances in the Synthesis and Electrocatalytic Applications of Platinum-Based Bimetallic Alloy Nanostructures. *Chemcatchem* **2015**, *7*, 3206–3228. [CrossRef]
8. Huang, X.Y.; Wang, A.J.; Zhang, L.; Fang, K.M.; Wu, L.J.; Feng, J.J. Melamine-assisted solvothermal synthesis of PtNi nanodentrites as highly efficient and durable electrocatalyst for hydrogen evolution reaction. *J. Colloid Interface Sci.* **2018**, *531*, 578–584. [CrossRef] [PubMed]
9. Tang, J.; Salunkhe, R.; Liu, J.; Torad, N.; Imura, M.; Furukawa, S.; Yamauchi, Y. Thermal Conversion of Core–Shell Metal–Organic Frameworks: A New Method for Selectively Functionalized Nanoporous Hybrid Carbon. *J. Am. Chem. Soc.* **2015**, *137*, 1572–1580. [CrossRef] [PubMed]
10. Wang, X.; He, J.; Yu, B.; Sun, B.; Yang, D.; Zhang, X.; Zhang, Q.; Zhang, W.; Gu, L.; Chen, Y. CoSe$_2$ nanoparticles embedded MOF-derived Co-N-C nanoflake arrays as efficient and stable electrocatalyst for hydrogen evolution reaction. *Appl. Catal. B Environ.* **2019**, *258*, 117996. [CrossRef]
11. Zhu, B.J.; Zou, R.; Xu, Q. Metal–Organic Framework Based Catalysts for Hydrogen Evolution. *Adv. Energy Mater.* **2018**, *8*, 1801193. [CrossRef]
12. Qin, Y.; Han, X.; Gadipelli, S.; Guo, J.; Wu, S.; Kang, L.; Callison, J.; Guo, Z. In situ synthesized low-PtCo@porous carbon catalyst for highly efficient hydrogen evolution. *J. Mater. Chem. A* **2019**, *7*, 6543–6551. [CrossRef]
13. Wang, Z.L.; Hao, X.-F.; Jiang, Z.; Sun, X.-P.; Xu, D.; Wang, J.; Zhong, H.-X.; Meng, F.-L.; Zhang, X.-B. C and N Hybrid Coordination Derived Co–C–N Complex as a Highly Efficient Electrocatalyst for Hydrogen Evolution Reaction. *J. Am. Chem. Soc.* **2015**, *137*, 15070–15073. [CrossRef]
14. Pan, F.; Zhang, H.; Liu, K.; Cullen, D.; More, K.L.; Wang, M.; Feng, Z.; Wang, G.; Wu, G.; Li, Y. Unveiling Active Sites of CO2 Reduction on Nitrogen-Coordinated and Atomically Dispersed Iron and Cobalt Catalysts. *ACS Catal.* **2018**, *8*, 3116–3122. [CrossRef]
15. Zhang, W.; Yao, X.; Zhou, S.; Li, X.; Li, L.; Yu, Z.; Gu, L. ZIF-8/ZIF-67-Derived Co-N-x-Embedded 1D Porous Carbon Nanofibers with Graphitic Carbon-Encased Co Nanoparticles as an Efficient Bifunctional Electrocatalyst. *Small* **2018**, *14*, 1800423. [CrossRef]
16. Pan, Y.; Sun, K.; Liu, S.; Cao, X.; Wu, K.; Cheong, W.-C.; Chen, Z.; Wang, Y.; Li, Y.; Liu, Y.; et al. Core–Shell ZIF-8@ZIF-67-Derived CoP Nanoparticle-Embedded N-Doped Carbon Nanotube Hollow Polyhedron for Efficient Overall Water Splitting. *J. Am. Chem. Soc.* **2018**, *140*, 2610–2618. [CrossRef] [PubMed]
17. A Han, S.; Lee, J.; Shim, K.; Lin, J.; Shahabuddin, M.; Lee, J.-W.; Kim, S.-W.; Park, M.-S.; Kim, J.H. Strategically Designed Zeolitic Imidazolate Frameworks for Controlling the Degree of Graphitization. *Bull. Chem. Soc. Jpn.* **2018**, *91*, 1474–1480. [CrossRef]
18. Li, H.; He, Y.; He, T.; Yan, S.; Ma, X.; Chen, J. ZIF-derived Co nanoparticle/N-doped CNTs composites embedded in N-doped carbon substrate as efficient electrocatalyst for hydrogen and oxygen evolution. *J. Mater. Sci. Mater. Electron.* **2019**, *30*, 21388–21397. [CrossRef]
19. Li, J.; Liu, L.; Ai, Y.; Hu, Z.; Xie, L.; Bao, H.; Wu, J.; Tian, H.; Guo, R.; Ren, S.; et al. Facile and Large-Scale Fabrication of Sub-3 nm PtNi Nanoparticles Supported on Porous Carbon Sheet: A Bifunctional Material for the Hydrogen Evolution Reaction and Hydrogenation. *Chem. A Eur. J.* **2019**, *25*, 7191–7200. [CrossRef] [PubMed]
20. Xu, M.; Chen, X.; Li, C.; Wang, T.; Zeng, Y.; Mu, S.; Yu, J. PtNi Supported on ZIF-Derived Porous Carbon as a High Efficiency Acidic Hydrogen Evolution Catalyst. *Energy Fuels* **2021**, *35*, 17861–17868. [CrossRef]
21. Chen, Y.; Jie, S.; Yang, C.; Liu, Z. Active and efficient Co-N/C catalysts derived from cobalt porphyrin for selective oxidation of alkylaromatics. *Appl. Surf. Sci.* **2017**, *419*, 98–106. [CrossRef]
22. Wang, N.; Li, L.; Zhao, D.; Kang, X.; Tang, Z.; Chen, S. Graphene Composites with Cobalt Sulfide: Efficient Trifunctional Electrocatalysts for Oxygen Reversible Catalysis and Hydrogen Production in the Same Electrolyte. *Small* **2017**, *13*, 1701025. [CrossRef] [PubMed]

23. Jiang, Y.; Wu, X.; Yan, Y.; Luo, S.; Li, X.; Huang, J.; Zhang, H.; Yang, D. Coupling PtNi Ultrathin Nanowires with MXenes for Boosting Electrocatalytic Hydrogen Evolution in Both Acidic and Alkaline Solutions. *Small* **2019**, *15*, 1805474. [CrossRef] [PubMed]

24. Chen, J.; Wang, J.; Chen, J.; Wang, L. A bifunctional electrocatalyst of PtNi nanoparticles immobilized on three-dimensional carbon nanofiber mats for efficient and stable water splitting in both acid and basic media. *J. Mater. Sci.* **2017**, *52*, 13064–13077. [CrossRef]

25. Wang, J.; Chen, J.; Zhu, H.; Zhang, M.; Du, M.L. Designed Synthesis of Size-Controlled Pt-Cu Alloy Nanoparticles Encapsulated in Carbon Nanofibers and Their High Efficient Electrocatalytic Activity Toward Hydrogen Evolution Reaction. *Adv. Mater. Interfaces* **2017**, *4*, 1700005. [CrossRef]

26. Yang, J.; Ning, G.; Yu, L.; Wang, Y.; Luan, C.; Fan, A.; Zhang, X.; Liu, Y.; Dong, Y.; Dai, X. Morphology controllable synthesis of PtNi concave nanocubes enclosed by high-index facets supported on porous graphene for enhanced hydrogen evolution reaction. *J. Mater. Chem. A* **2019**, *7*, 17790–17796. [CrossRef]

27. Wu, Q.; Jin, H.; Chen, W.; Huo, S.; Chen, X.; Su, X.; Wang, H.; Wang, J. Graphitized nitrogen-doped porous carbon composites derived from ZIF-8 as efficient microwave absorption materials. *Mater. Res. Express* **2018**, *5*, 065602. [CrossRef]

28. Xie, L.; Liu, Q.; Shi, X.; Asiri, A.M.; Luo, Y.; Sun, X. Superior alkaline hydrogen evolution electrocatalysis enabled by an ultrafine PtNi nanoparticle-decorated Ni nanoarray with ultralow Pt loading. *Inorg. Chem. Front.* **2018**, *5*, 1365–1369. [CrossRef]

29. Zhou, S.; Chen, X.; Yu, P.; Gao, F.; Mao, L. Nitrogen-doped carbon nanotubes as an excellent substrate for electroless deposition of Pd nanoparticles with a high efficiency toward the hydrogen evolution reaction. *Electrochem. Commun.* **2018**, *90*, 91–95. [CrossRef]

30. Devadas, B.; Imae, T. Hydrogen evolution reaction efficiency by low loading of platinum nanoparticles protected by dendrimers on carbon materials. *Electrochem. Commun.* **2016**, *72*, 135–139. [CrossRef]

31. Chen, X.; Xu, M.; Li, S.; Li, C.; Sun, X.; Mu, S.; Yu, J. Ultrafine IrNi Bimetals Encapsulated in Zeolitic Imidazolate Frameworks-Derived Porous N-Doped Carbon for Boosting Oxygen Evolution in Both Alkaline and Acidic Electrolytes. *Adv. Mater. Interfaces* **2020**, *7*, 2001145. [CrossRef]

Article

Non-PGM Electrocatalysts for PEM Fuel Cells: A DFT Study on the Effects of Fluorination of FeN$_x$-Doped and N-Doped Carbon Catalysts

Mohamed Cherif, Jean-Pol Dodelet, Gaixia Zhang , Vassili P. Glibin, Shuhui Sun and François Vidal *

Centre Énergie, Matériaux, Télécommunications, Institut National de la Recherche Scientifique, 1650 Bd. Lionel-Boulet, Varennes, QC J3X 1S2, Canada; mohamed.cherif@inrs.ca (M.C.); jean-pol.dodelet@inrs.ca (J.-P.D.); gaixia.zhang@inrs.ca (G.Z.); vassili.glibin@gmail.com (V.P.G.); shuhui.sun@inrs.ca (S.S.)
* Correspondence: francois.vidal@inrs.ca

Abstract: Fluorination is considered as a means of reducing the degradation of Fe/N/C, a highly active FeN$_x$-doped disorganized carbon catalyst for the oxygen reduction reaction (ORR) in PEM fuel cells. Our recent experiments have, however, revealed that fluorination poisons the FeN$_x$ moiety of the Fe/N/C catalytic site, considerably reducing the activity of the resulting catalyst to that of carbon only doped with nitrogen. Using the density functional theory (DFT), we clarify in this work the mechanisms by which fluorine interacts with the catalyst. We studied 10 possible FeN$_x$ site configurations as well as 2 metal-free sites in the absence or presence of fluorine molecules and atoms. When the FeN$_x$ moiety is located on a single graphene layer accessible on both sides, we found that fluorine binds strongly to Fe but that two F atoms, one on each side of the FeN$_x$ plane, are necessary to completely inhibit the catalytic activity of the FeN$_x$ sites. When considering the more realistic model of a stack of graphene layers, only one F atom is needed to poison the FeN$_x$ moiety on the top layer since ORR hardly takes place between carbon layers. We also found that metal-free catalytic N-sites are immune to poisoning by fluorination, in accordance with our experiments. Finally, we explain how most of the catalytic activity can be recovered by heating to 900 °C after fluorination. This research helps to clarify the role of metallic sites compared to non-metallic ones upon the fluorination of FeN$_x$-doped disorganized carbon catalysts.

Keywords: oxygen reduction reaction; proton exchange membrane fuel cell; fluorination; density functional theory; non-noble metal catalyst; N-doped carbon catalyst

Citation: Cherif, M.; Dodelet, J.-P.; Zhang, G.; Glibin, V.P.; Sun, S.; Vidal, F. Non-PGM Electrocatalysts for PEM Fuel Cells: A DFT Study on the Effects of Fluorination of FeN$_x$-Doped and N-Doped Carbon Catalysts. *Molecules* **2021**, *26*, 7370. https://doi.org/10.3390/molecules26237370

Academic Editors: Jingqi Guan and Yin Wang

Received: 26 October 2021
Accepted: 1 December 2021
Published: 4 December 2021

Publisher's Note: MDPI stays neutral with regard to jurisdictional claims in published maps and institutional affiliations.

1. Introduction

While promising non-noble metal catalysts for the oxygen reduction reaction (ORR) in proton-exchange membrane (PEM) hydrogen fuel cells have been synthesized over the years [1–5], their stability in fuel cells remains the main obstacle to their widespread use [6,7]. One of the most promising non-noble metal catalysts synthesized to date is FeN$_x$-doped disorganized carbon [8–11]. There are several experimental and theoretical pieces of evidence that the Fe atom is the site where the ORR takes place [11–19]. It has been observed that this type of catalyst suffers from a decrease of almost half of its activity in a few hours of operation in fuel cells, followed by a much slower decrease thereafter. The current delivered by the fuel cell versus time can be fitted by a double exponential decay [20]. Few hypotheses have been put forward to explain the first rapid decay of catalytic activity. These include demetallation of the metal catalytic sites [20–23] and chemical reactions with H$_2$O$_2$ [24–27]. The slower decay has not attracted as much interest as the fast one to date. Recent simulation work suggests that planar M$_3$(C$_6$O$_6$)$_2$ [28] and M$_3$(C$_6$S$_3$O$_3$)$_2$ [29] structures, where M is a transition metal, may also be promising candidates but these have not yet passed the test of experiment.

There are indications that the fluorination of materials in acidic media improves their oxidative stability. Examples of such systems include Nafion ionomer, Pt/C, and platinum group metal-free catalysts for PEM fuel cells [30–32]. Recently, we also fluorinated a highly active FeN_x-doped carbon catalyst in the hopes that fluorine would increase its stability in PEM fuel cells [33]. However, even after a short (2 min) exposure to a room-temperature F_2:N_2 (1:1; *vol.*) gas stream, fluorination considerably inhibited the catalyst performance in H_2/O_2 PEM fuel cells. Even if these experiments did not yield the expected results, they enabled several important observations to be made regarding the properties of the catalytic material under study:

(1) The catalytic activity of the fluorinated Fe/N/C catalyst became similar to that of Fe-free nitrogen-doped carbon catalysts;

(2) The XPS F1s spectra revealed that most Fe sites were associated with a single F atom and fewer were associated with two F atoms;

(3) A total of 70% of the initial activity could be recovered after a heat treatment of the F-poisoned catalyst at 900 °C in Ar.

The observations made in the context of these fluorination experiments have in fact provided a unique opportunity to improve our understanding of the nature of our FeN_x-doped carbon catalysts and of the decay mechanisms of their catalytic activity in PEM fuel cells. In order to support and deepen the conclusions of our experimental study, in this paper we report density functional theory (DFT) calculations, based on the current understanding of the atomic structure of the catalytic sites and processes, and study the catalytic properties of these sites in the absence/presence of adsorbed fluorine.

Several theoretical studies have already focused on MN_x-doped carbon catalysts, most often Fe [34–38], Co [35], Mn [35,39], and Ni [35]. Per the indications of several experimental studies [34,39–41], they conclude that the catalytic site is, specifically, the M atom within a functional group MN_x embedded in a planar carbon layer. It is generally thought that the ORR catalyzed on these sites follows the four-electron exchange process [42–45]

$$
\begin{aligned}
* + O_2 + 4(H^+ + e^-) \quad &\rightarrow \quad *O_2 + 4(H^+ + e^-) \quad \rightarrow \\
\text{I} \qquad\qquad &\qquad\qquad \text{II} \\
* OOH + 3(H^+ + e^-) \quad &\rightarrow \quad *O + H_2O + 2(H^+ + e^-) \quad \rightarrow \\
\text{III} \qquad\qquad &\qquad\qquad \text{IV} \\
*OH + H_2O + (H^+ + e^-) \quad &\rightarrow \quad * + 2H_2O \\
\text{V} \qquad\qquad &\qquad\qquad \text{VI}
\end{aligned}
\tag{1}
$$

where * denotes the adsorption site and the labels I to VI refer to the six reaction steps.

For several MN_x-doped carbon structures, it has been found that, at low enough potentials, the free energy of each step of the reaction sequence (1) decreases uniformly from the first to the last step, indicating that the reaction sequence (1) is thermodynamically viable at these potentials. Other possible pathways, such as those involving spontaneous O_2 dissociation or H_2O_2 formation, are less likely due to the increase in free energy at some stage of the process [24]. Several DFT studies have also been carried out for catalysts without metal [45–51]. These generally consider N-doped carbon structures and assume that the reaction sequence (1) still takes place at low enough potentials. These catalysts appear to be thermodynamically viable for some carbon sites near a nitrogen atom. However, O_2 adsorbs weakly or not at all on the catalytic sites (step II in the sequence (1)). This characteristic likely explains, at least in part, the much lower activity of these sites compared to the higher activity obtained with metal sites.

In a recent work, we theoretically studied the fluorination of two single-layer porous FeN_4-doped carbon structures, one with pyrrolic nitrogen atoms and the other with pyridinic nitrogen atoms at the FeN_4 sites, and we assumed that the catalytic reaction took place through the sequence (1) [52]. Subsequent work has investigated ORR for various adsorbates bound to several transition metals on an MN_4 site [53,54]. However, actual catalysts most likely contain many embodiments of MN_x moieties as well as M-free N atoms in carbon layers, and involve more than one carbon layer. In order to provide a more complete picture of the fluorination process and of its influence on ORR, we performed DFT calculations for nine additional atomic structures of FeN$_x$-doped carbon sites with x between 1 and 4 as well as for two N-doped metal-free carbon structures. We also examined the possibility of whether the ORR can be catalyzed on an Fe site between two parallel carbon layers in the presence of a F atom bound on Fe on the opposite side.

It is generally believed that the catalytic sites for the ORR of the FeN$_x$-doped carbon materials consist of planar FeN$_x$ moieties located in a carbon structure which is generally approximated by a single carbon layer. Figure 1 shows some of the possible embodiments of such structures. Of course, the set of structures shown in Figure 1 is not exhaustive. Many variants of each structure are possible, such as, for example, pores in the carbon layer (as in Figure 1j), the FeN$_x$ moiety being near the edge of the carbon layer (as in Figure 1g,i), and N atoms being randomly distributed in the carbon layer (as in Figure 1e). Also, the N atoms surrounding the Fe ion may be either of a pyridinic type (Figure 1a–i) or of a pyrrolic type (Figure 1j). We will consider the 10 structures shown in Figure 1, expecting that they will be representative of the main effects of fluorination on the properties of the catalysts. For the purpose of the following discussion, special attention will be paid to the structure in Figure 1j. For this structure, the F–N bond length is 2.00 Å and the N–Fe–N angles are 197.19° and 82.81°.

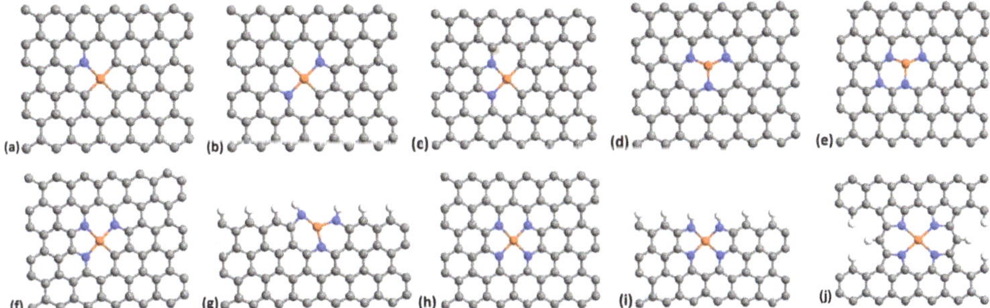

Figure 1. DFT optimized configurations of FeN$_x$-doped carbon catalysts investigated in this work. Panels (**a**–**j**) refer to the 10 configurations considered in this work. Color code: grey is carbon, blue is nitrogen, orange is iron.

It can be found in the literature that the free energy at zero potential of the steps of the reaction sequence (1) for the pristine structures of Figure 1c [55], 1h [56], and 1j [52] is uniformly descending at each reaction step (see Section 2.1). The structure of Figure 1b has also been investigated and was found to be not thermodynamically viable [55] (because the free energy presents a minimum at step V of the reaction sequence (1)—see Section 2.1). The other sites shown in Figure 1 have, to the best of our knowledge, never been investigated so far.

Figure 2 shows the two metal-free nitrogen-doped carbon catalytic sites considered in this work. As reported above, the activity of the fluorinated Fe/N/C catalysts became similar to that of Fe-free nitrogen-doped carbon catalysts. It is assumed that the reaction sequence is given by (1) where * now denotes an active carbon site. Other nitrogen-doped carbon structures have been investigated and shown to be potential catalysts for ORR [16–49]. The structures shown in Figure 2 were selected for this work because they turn out to have the lowest formation energies [45] and are, therefore, most likely to be found in actual catalysts. Only two of these structures are considered in this work because we show that their catalytic properties are immune to fluorination and this result is sufficient for our purpose.

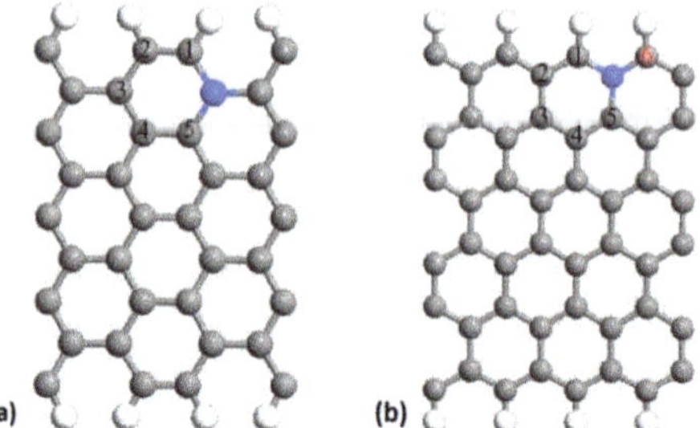

Figure 2. DFT optimized configurations of metal-free catalysts investigated in this work. (**a**) Armchair configuration of N-doped carbon and (**b**) zigzag configuration of N-doped carbon. The meaning of the numbers on the carbon atoms is discussed in Section 2.3.

2. Results

2.1. Fluorination of the FeN$_x$ Sites—Single Carbon Layer

We first performed DFT optimizations of all the non-fluorinated carbon-based catalysts considered in this work. The resulting structures are shown in Figures 1 and 2. We then introduced F atoms and F$_2$ molecules at different locations on these atomic structures and performed new DFT optimizations.

We first considered the adsorption of F$_2$ on the Fe atom of the FeN$_x$ sites. We found that F$_2$ adsorbs on Fe for all the FeN$_x$ configurations of Figure 1. When adsorbed on Fe, the F$_2$ molecule is strongly stretched relative to the free F$_2$ molecule. In the example of Figure 3a, where we use the basic structure of Figure 1j, the spacing between the two F atoms of the adsorbed F$_2$ is 2.32 Å while the spacing of the two F atoms in the free F$_2$ molecule is 1.11 Å. In fact, the adsorbed F$_2$ molecule is subject to dissociation because, for example, the binding energy of the dissociated F$_2$ molecule with one F atom adsorbed on Fe and the other F atom adsorbed on a near C atom either on the same side (Figure 3b) or on the opposite side (Figure 3c) of the carbon layer is, respectively, 1.36 eV and 1.40 eV lower. The fluorinated structure is considerably more stable when two F atoms are adsorbed on the Fe atom on opposite sides of the carbon layer (Figure 3d) because the energy of the system is 6.12 eV lower than that of F$_2$ adsorbed on Fe (Figure 1a). The binding energies of the fluorine adsorbates shown in Figure 3 are given in Table 1. All F–Fe and F–C bond lengths appearing in Figure 3 are shown in Table A1 of the Appendix A.

In the case where two F atoms are adsorbed on Fe on both sides of the carbon layer (Figure 3d), the Fe site becomes unavailable to catalyze ORR. Indeed, the binding energy of the O$_2$ molecule on the Fe atom of the pristine structure of Figure 3 is only −0.71 eV, meaning that the fluorinated iron sites of Figure 3d,e are very stable and the adsorbed F is unlikely to be spontaneously replaced by O$_2$. The same conclusion holds for all structures of Figure 1, except for Figure 1d, for which the binding energy of Fe–O$_2$ is stronger than that of Fe–F, as shown in Table 2. However, this structure appears to be inactive in the absence of fluorine, as can be seen in Figure 4d.

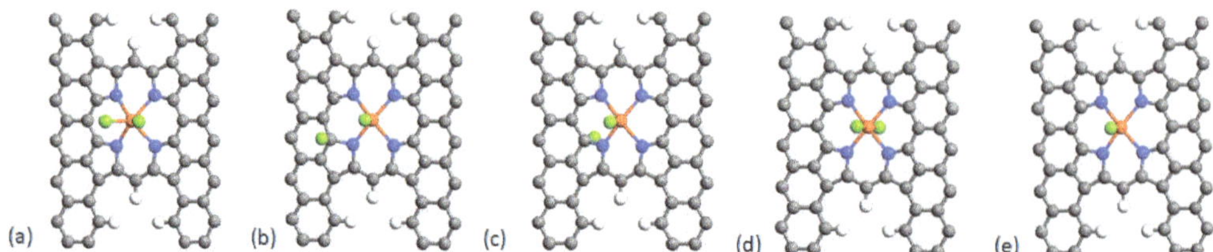

Figure 3. Adsorption of F$_2$ and F on the Fe site for the basic structure of Figure 1j. (**a**) Adsorption of F$_2$; (**b,c**) adsorption of one F on Fe and one F on a nearby carbon site; (**d**) adsorption of two F on the Fe site on both sides of the carbon layer; and (**e**) adsorption of a single F on Fe. Color code: grey is carbon, blue is nitrogen, orange is iron, and green is fluorine.

Table 1. Binding energies of the fluorine adsorbates for the structures shown in Figure 3.

Structure	Binding Energy (eV)
a	−1.76
b	−3.12
c	−3.16
d	−7.88
e	−4.56

Table 2. Binding energies in eV for F adsorbed on Fe (Fe–F), for two F adsorbed on Fe on both sides of the carbon layer (F–Fe–F), and for O_2 adsorbed on Fe (Fe–O_2) for the structures shown in Figure 1.

Structure	Fe–F	F–Fe–F	Fe–O_2
a	−4.27	−12.10	−0.84
b	−4.00	−11.92	−2.50
c	−3.85	−11.88	−2.06
d	−3.94	−13.04	−4.60
e	−4.05	−12.02	−1.82
f	−4.66	−12.96	−3.50
g	−5.89	−12.94	−4.55
h	−3.18	−11.88	−1.37
i	−2.93	−10.40	−1.04
j	−4.56	−7.88	−0.71

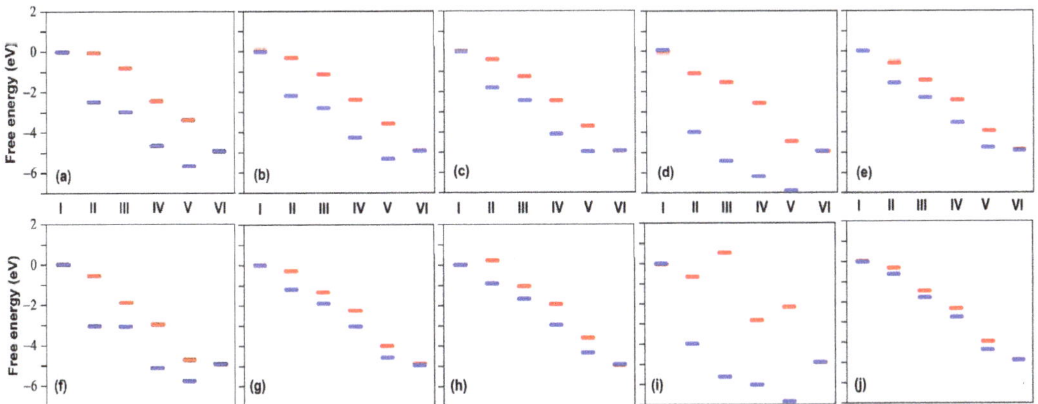

Figure 4. Relative free energy at zero potential for the six steps (I–VI) of the ORR sequence (1) for the atomic structures shown in Figure 1 with (red segments) and without (blue segments) a F atom adsorbed on Fe. The panels (**a**–**j**) correspond to the structures shown in Figure 1a to j.

For a fluorinated FeN_x site located in a single carbon layer, the adsorption of a single F atom on the Fe site can turn a poor catalyst into an effective one due to the weakening of the binding energy of the adsorbates of the reaction sequence (1). This can be seen in Figure 4, which shows the relative free energies (with respect to the initial state) of each step of the catalytic reaction sequence (1) in the cases where the Fe sites are free of fluorine (blue segments) and where a F atom is adsorbed on Fe (red segments). One can see for the structures corresponding to Figure 4a,b,d,f, that the ORR catalytic process, which seems unlikely without the adsorption of F because of the very low energy level of step V, $*OH + H_2O + (H^+ + e^-)$, becomes thermodynamically viable with the adsorption of F on Fe. However, the structure corresponding to Figure 4i remains a poor catalyst with and without the adsorption of F on Fe. On the other hand, the structure corresponding to Figure 4h, which seems to be a possible good catalyst without fluorine, becomes less effective with fluorine because O_2 can no longer adsorb on Fe (step II). Table A2 shows the F–Fe, O–Fe–O, and O–O bond lengths in O_2, Fe–O_2, and F–Fe–O_2. It can be seen that the F–Fe bond length in F–Fe–O_2 is almost the same as in Figure 3e (Table A1). However, the Fe–O and O–O bond lengths in F–Fe–O_2 are slightly smaller than in Fe–O_2, indicating that the Fe–O_2 bond is weakened due to the presence of F, in agreement with the discussion above.

From this, we conclude that carbon-based catalysts with sufficient separation between carbon layers (which is equivalent to considering the catalytic sites located on a single carbon layer) could theoretically benefit from partial fluorination by the weakening of the adsorbate free energy. This effect leads to a more favorable free energy distribution for most of the fluorinated ORR active sites as compared to that of the sites without any F–Fe bond. The conclusion that ORR is generally promoted when an adsorbate is bound to the metal atom on the opposite side of the carbon plane was also obtained via the DFT calculations reported in [53,54].

2.2. Fluorination of the FeN_x Sites—Double Carbon Layer

The useful FeN_x active sites (those able to produce ORR) are actually thought to be mostly embedded at the surface of continuous graphene layer stacks or located between two discontinuous graphene layers of micro- or mesopores [20]. Therefore, the question naturally arises as to whether the ORR can take place between the first two carbon layers when a F atom is adsorbed on Fe on the more accessible free side of the FeN_x site. To answer this question, we considered a model using two carbon layers: on the top, a carbon layer containing the FeN_x moiety and, under this first layer, another one composed of a single graphene layer parallel to the first one and located at 3.6 Å from the first layer. We selected this interplanar distance, which is a little larger than that of d = 3.35 Å in graphite, because carbon is disorganized after the pyrolysis stage. As a matter of fact, a fairly broad distribution of d-spacings (between 3.5 and 4.1 Å) was measured for furnace turbostratic carbon black grades, regardless of their particle size and structure. The average TEM measured d-spacings range between 3.83 and 3.92 Å and are significantly larger than the X-ray measured d-spacings ranging from 3.52 to 3.56 Å [57]. Thus, we used the intermediate value of 3.6 Å.

The first step in ORR is the adsorption of O_2 on Fe. Then, according to the reaction sequence (1), each adsorbate combines with an H^+ ion and an electron. Therefore, O_2 and H^+ have to migrate between the carbon layers to reach the Fe atom of the FeN_x site. This migration is certainly easier for porous carbon structures such as the one of Figure 1j or when the Fe atom is close to the edge of the carbon layer, as in Figure 1g,i. Figure 5 shows the basic structure of Figure 3e to which a parallel graphene plane was added under the graphene plane containing the F–FeN_x site. One notes that there is apparently enough room between the two carbon layers to accommodate an oxygen atom or molecule because the spacing between the carbon layers of our amorphous carbon catalyst is assumed here to be 3.6 Å, while the theoretical radii of Fe, C, and O are 1.56 Å, 0.67 Å, and 0.48 Å [58], respectively, so that the sum of the radius of Fe, the diameter of O_2, and the radius of C is 3.19 Å, which is smaller than the assumed spacing between the carbon layers.

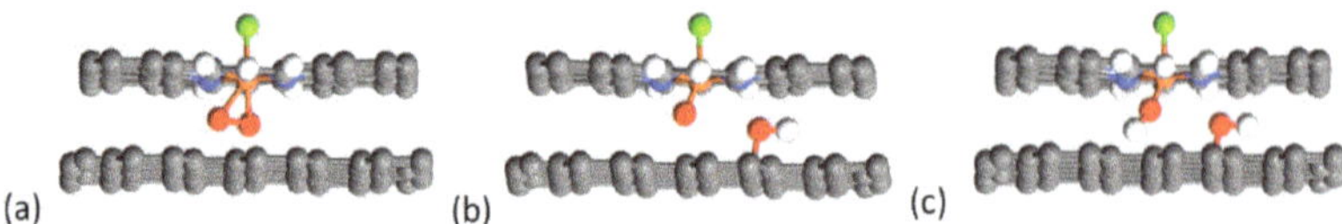

Figure 5. The first catalytic reaction steps between two carbon layers with a F atom adsorbed on Fe on the free side for the basic structure of Figure 1j. (**a**) Adsorption of O_2 on Fe; (**b**) result of the spontaneous dissociation of OOH into O on Fe and OH on the opposite carbon layer; and (**c**) formation of OH adsorbed on Fe between the two layers. Color code: grey is carbon, blue is nitrogen, orange is iron, green is fluorine, and red is oxygen.

The free energy diagram of the catalytic reaction is shown in Figure 6. O_2 can adsorb on the Fe atom between the planes (Figure 5a and step II in Figure 6) but it needs around 1 eV to get there. This is a consequence of the stress induced on the surrounding structure by the insertion of the O_2 molecule. Our transition state calculation indicates that the activation energy is around 1.5 eV between the state of O_2 in the pore and its adsorbed state on the Fe atom. However, from the latter adsorbed state (step II in Figure 6), a continuously decreasing free energy sequence can be found, but with modifications with respect to the sequence (1). When adding $H^+ + e^-$ to the adsorbed O_2, OOH spontaneously dissociates into an O adsorbed on Fe and an OH which adsorbs on a carbon atom of the opposite layer (Figure 5b and step III in Figure 6). The reactions up to step III in Figure 6 are as follows:

$$Fe + G + O_2 + 4(H^+ + e^-) \quad \rightarrow \quad (Fe - O_2) + G + 4(H^+ + e^-) \quad \rightarrow \quad (Fe - O) + (G - OH) + 3(H^+ + e^-) \qquad (2)$$
$$\quad\quad\quad I \quad\quad\quad\quad\quad\quad\quad\quad\quad\quad\quad\quad II \quad\quad\quad\quad\quad\quad\quad\quad\quad\quad\quad\quad III$$

where G is the bottom graphene plane. Then three distinct paths are possible, depending on how the successive $H^+ + e^-$ are added to the adsorbates. These paths are illustrated in Figure 6 by black, blue, and red segments, respectively. The last step (VI) is $Fe + G + 2H_2O$, i.e., the formation of two water molecules after the exchange of four electrons. The steps IV and V for the three paths are, respectively,

$$Path\ 1: \ (Fe - O) + G + H_2O + 2(H^+ + e^-) \quad \rightarrow \quad (Fe - OH) + G + H_2O + (H^+ + e^-) \qquad (3)$$
$$\quad\quad\quad\quad\quad\quad IV \quad\quad\quad\quad\quad\quad\quad\quad\quad\quad\quad\quad\quad V$$

$$Path\ 2: \ (Fe - OH) + (G - OH) + 2(H^+ + e^-) \quad \rightarrow \quad Fe + (G - OH) + H_2O + (H^+ + e^-) \qquad (4)$$
$$\quad\quad\quad\quad\quad\quad IV \quad\quad\quad\quad\quad\quad\quad\quad\quad\quad\quad\quad\quad V$$

$$Path\ 3: \ (Fe - OH) + (G - OH) + 2(H^+ + e^-) \quad \rightarrow \quad (Fe - OH) + G + H_2O + (H^+ + e^-) \qquad (5)$$
$$\quad\quad\quad\quad\quad\quad IV \quad\quad\quad\quad\quad\quad\quad\quad\quad\quad\quad\quad\quad V$$

The most complex intermediate state, where one OH is adsorbed on Fe and the other OH is adsorbed on the opposite graphene layer G, is shown in Figure 5c and corresponds to step IV of paths 2 and 3.

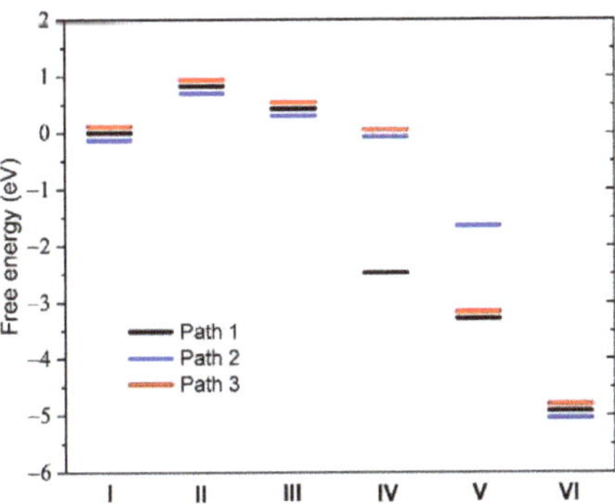

Figure 6. Relative free energy at zero potential for the six steps of the ORR sequences between two carbon layers for the structure of Figure 3e to which a parallel graphene plane was added under the graphene plane containing the F–FeN$_x$ site. The three possible paths are explained in the text. The energy levels have been shifted slightly to facilitate path identification.

Since the theoretical radius of F (0.42 Å) is smaller than that of O (0.48 Å) [58], F and F_2 can also be accommodated between the two layers. We also performed DFT calculations for that case. The results are illustrated in Figure 7. The binding energies of F_2 and F in the cases of Figure 7a,b are −1.84 and −4.48 eV, respectively. Here the reference structure is composed of the two planes with the external adsorbed F atom on Fe. Thus, as in the monolayer case of Figure 3, the dissociation of F_2 is favored between the two layers of carbon. In addition, contrary to O_2, the adsorption of F_2 (or its dissociated form) between the plane is exothermic. For the sake of comparison with the single plane case of Figure 3d, the binding energy of the two F on both sides of the plane is -6.24 eV (Figure 7c) vs. -7.88 eV in the case of Figure 3d. Again, the difference in binding energy is due to the stress induced on the surrounding structure by the insertion of the F atom. The existence of F–Fe–F bonds in the catalyst could correspond to the specific peak at ~685.4 eV assigned to the adsorption of two F atoms on Fe in the F1s XPS spectrum of the catalyst [33].

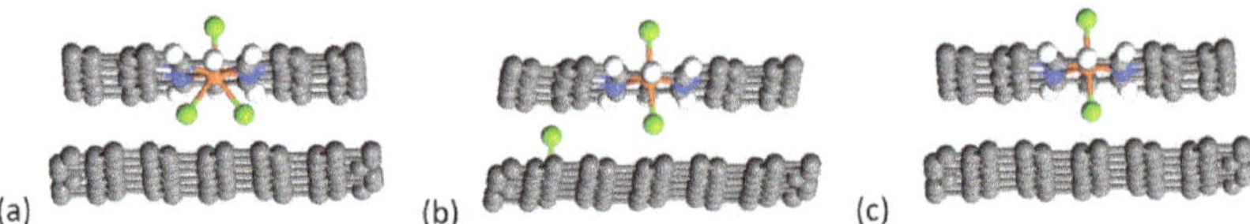

Figure 7. Adsorption of F_2 and F on the Fe site between two carbon layers for the basic structure of Figure 1j. (**a**) Adsorption F_2 on Fe; (**b**) adsorption of F on Fe and F on a nearby carbon site (dissociated form of F_2); and (**c**) adsorption of F on Fe between two carbon layers.

From this study of the fluorinated bilayer configuration, it appears that ORR catalysis between two carbon layers is inefficient, primarily because O_2 requires about 1 eV to occupy the catalytic site, although the subsequent reaction steps are thermodynamically viable. Furthermore, upon fluorination of the catalyst, F_2 and F can occupy the catalytic site at a lower energy cost, thereby poisoning the FeN_x sites.

2.3. Fluorination of Metal-Free Sites

We now turn to the metal-free catalysts shown in Figure 2, which are also known to contribute to the ORR, but to a much lesser extent than the FeN_x metal sites free of fluorine [59]. It has been demonstrated that these catalysts can produce uniformly descending free energy steps for the reaction sequence (1), as for some of the FeN_x sites (see Figure 4). However, O_2 hardly adsorbs on these structures [45–49]. Our DFT calculations are in good agreement with those previous works, as shown by the blue segments in Figure 8a,b, which were obtained by considering the ORR active sites 5 and 1 in Figure 2a,b, respectively.

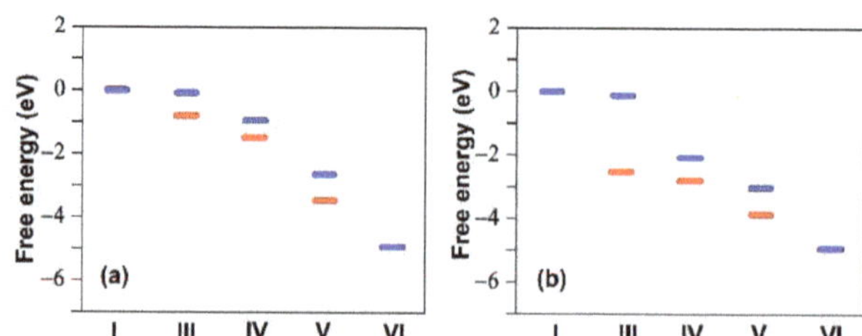

Figure 8. Relative free energy at zero potential for the six steps (I–VI) of the ORR sequence (1) for the sites shown in Figure 2 with (red segments) and without (blue segments) a F atom adsorbed on a carbon atom. (**a**) Armchair configuration of N-doped carbon; (**b**) zigzag configuration of N-doped carbon. The blue segments in (a) were obtained by considering the ORR activity of site 5 in Figure 2a. The blue segments in (b) were obtained by considering the ORR activity of site 1 in Figure 2b. The red segments in (a) were obtained by considering a F atom adsorbed on site 1 in Figure 2a, while keeping site 5 as ORR active. The red segments in (b) were obtained by considering a F atom adsorbed on site 1 in Figure 2b, while considering the ORR activity of site 6.

We then verified whether the carbon catalytic sites can be poisoned by F or F_2. For the armchair configuration of Figure 2a, adsorption of F on the sites numbered from 1 to 5 were tested. Table 3 shows that F can adsorb on the five sites with different binding energies.

Site 1 has the strongest binding energy of -2.57 eV. When a F atom is adsorbed on site 1, Figure 8a (red segments) shows that the catalytic site 5 remains active.

Table 3. Binding energies in eV for a F atom adsorbed on the 5 carbon sites of Figure 2a (armchair) and Figure 2b (zigzag).

Structure	Site 1	Site 2	Site 3	Site 4	Site 5
Armchair	-2.57	-2.12	-2.33	-1.69	-2.18
Zigzag	-3.78	-3.52	-2.80	-2.36	-2.97

For the zigzag structure we found that the catalytic site 1 in Figure 2b has the strongest binding energy of -3.78 eV for F. In this case, we carried out a calculation of the catalytic sequence on site 6 in Figure 2b. Even if the active site 1 is occupied by a fluorine atom, the ORR can take place on site 6, as can be seen in Figure 8b (red segments). Additional F–C bond length data for the five sites are given in Table A3. We note the anti-correlation between F–C bond length and binding energy as well as the smaller values compared to the F–Fe bond lengths given in Table A1.

We were unable to find a site where F_2 could be adsorbed on both the armchair and zigzag structures, so that dissociation of F_2 is unlikely on these structures. Therefore, F_2 has to be dissociated elsewhere, such as on the Fe sites, for instance, or at any oxygenated functionality like COH, COOH, or C–O–C, known (by XPS) to be present at the surface of our (and many other) non-PGM catalysts [33]. These calculations tend to confirm that fluorination does not affect much the catalytic activity of the metal-free catalysts considered here, thus providing an explanation for the residual ORR catalytic activity found after fluorination up to a value of F/C = 0.27 (measured by NMR) of the fluorinated Fe/N/C catalyst observed in the experiments reported in [33].

2.4. Defluorination of the FeN_x Sites

Here we consider a question that has puzzled us for some time: the possibility to thermally de-fluorinate at 900 °C previously fluorinated FeN_x catalytic sites such as the ones illustrated, for instance, in Figure 3d,e. Let us label these configurations F–FeN_4–F and F–FeN_4, respectively. From Table 1, the binding energies of the fluorine adsorbates on the Fe atom are -4.56 eV for F–FeN_4 and -7.88 eV for F–FeN_4–F. Despite the high binding energies of these bonds, it was experimentally found that the latter are broken after a 30 min heat treatment at 900 °C in Ar of the fluorinated catalysts (no F1s XPS signal anymore, as seen in Figure 6A of [33]). How is this possible when the thermal energy at 900 °C is only around 0.1 eV?

It is certainly not because of the special nature of the catalytic sites such as those illustrated in Figure 3e,d, as Table 2 confirms that the FeN_x sites illustrated in Figure 1 are all characterized by Fe–F and F–Fe–F binding energies of several eV. De-fluorination is only possible if the fluorinated FeN_x sites are involved in reactions also involving either radicals or small molecules released from the catalyst surface under heat treatment. Reactions (6–9) below are examples of possible de-fluorination reactions. Our thermodynamic calculations show that all these reactions are characterized by a negative free energy change ΔG at 900 °C, meaning that they are spontaneous at that temperature.

$$(F-FeN_4-F)_{solid} + (CH\bullet)_{gas} \leftrightarrow (F-FeN_4)_{solid} + C_{solid\ (graphene)} + (HF)_{gas} \quad (6)$$

$$(F-FeN_4)_{solid} + (CH\bullet)_{gas} \leftrightarrow (FeN_4)_{solid} + C_{solid\ (graphene)} + (HF)_{gas} \quad (7)$$

$$(F-FeN_4-F)_{solid} + (CF\bullet)_{gas} \leftrightarrow (F-FeN_4)_{solid} + (CF_2\bullet)_{gas} \quad (8)$$

$$2(F-FeN_4-F)_{solid} + (C_2F_6)_{gas} \leftrightarrow 2(F-FeN_4)_{solid} + 2(CF_4)_{gas} \quad (9)$$

In these examples, $(CH\bullet)_{gas}$, $(CF\bullet)_{gas}$, and $(C_2F_6)_{gas}$ are decomposition products [60] generated at 900 °C from the carbonaceous or from the fluorinated carbonaceous supports of the catalysts. The evidence for the release of such gases is documented by the TGA curves already reported in several figures of [33] for these fluorinated catalysts.

3. Computational Methods

All DFT calculations reported here were done using the Vienna ab initio software package (VASP) [61–64]. The calculations were performed using the generalized gradient approximation (GGA) with the Perdew–Burke–Ernzerhof (PBE) functional [65]. The convergence criterion on the relative energy was set to 10^{-5} and the plane wave energy cut-off was set to 500 eV for all calculations. The Brillouin zone was sampled on regular $4 \times 4 \times 4$ gamma grids. A graphene sheet with cell dimensions of a = 20.22 Å and b = 14.88 Å was used as a model for the carbon support. A void of 15 Å was included in the normal direction to avoid interactions between the periodic FeN_x-doped carbon layers. The doped carbon structures were created by substituting carbon atoms of the graphene sheet by FeN_x groups or by N atoms in the case of metal-free catalysts. The positions of all atoms were fully relaxed, except in the case of the two carbon layers, where the positions of the carbon atoms were fixed to prevent the planes from moving relative to each other. However, for the calculation of the activation energy of O_2 transiting between the two planes, in relation to Figure 5a, we used constraints where the edges of the planes were fixed along the x-axis while keeping the edges along the y-axis fixed, and vice versa. The activation energy was almost the same (1.5 eV) in both cases. The binding energy of an adsorbate on a given site was calculated using

$$E_{catalyst+adsorbate} - \left(E_{\text{catalyst}} + E_{adsorbate} \right) \tag{10}$$

where $E_{catalyst+adsorbate}$ is the energy of the carbon-doped catalyst with the adsorbate, $E_{catalyst}$ is the energy of the catalyst alone, and $E_{adsorbate}$ is the energy of the adsorbate far from the catalyst. The energy of $H^+ + e^-$ is taken as half the energy of the H_2 molecule, since H_2 is at an equilibrium with its dissociated form $2\left(H^+ + e^-\right)$ at the anode [43]. The molecules of the gas phases considered in this work, namely O_2, H_2, and F_2, are assumed to be non-interacting with each other, which implies that only single molecules have been considered. Each step of the catalytic sequence (1) corresponds to a free energy given by

$$G = G_0 + ZPE + TdS + G_{\text{sol}} \tag{11}$$

where G_0 is the energy of the structure per cell, ZPE is the zero point energy, TdS is the entropy term, and G_{sol} is the solvation energy arising by the aqueous medium. As was done in some of our previous works [52,54], for simplicity we assumed that the sum of the last three contributions nearly cancels, in agreement with [43,66]. However, corrections were brought to the intermediate state $*O + H_2O + 2\left(H^+ + e^-\right)$ and $* + 2H_2O$ of the reaction sequence (1), which were inferred to be +0.4 and −0.6 eV, respectively [43].

4. Conclusions

We used DFT to examine the consequences of fluorination of the FeN_x-doped and N-doped carbon catalysts used for ORR at the cathode of H_2/O_2 fuel cells. The main objectives of these calculations were to rationalize some of the experimental observations and to verify our conceptual representation of the catalytic sites and processes. We have considered several moieties of catalytic sites of FeN_x-doped carbon with x ranging from 1 to 4. Most of them seem to be suitable catalysts for ORR because the free energy of the supposed catalytic sequence decreases regularly at zero potential. When the FeN_x sites are located on a single graphene layer, it turns out that F_2 binds to Fe at FeN_x sites, with a binding energy of approximately −2 eV, but is subject to dissociation, leaving a single F on Fe with a binding energy of approximately −4 eV, which is stronger than the typical binding energy of O_2 on Fe. In these conditions, ORR cannot happen on the F-poisoned FeN_x side, but is still possible on the other side of the F–FeN_x site, even transforming some otherwise poor un-poisoned FeN_x catalytic configurations into better F–FeN_x active ones. In addition, two F atoms can also bind to Fe on both sides of the carbon layer with almost twice the binding energy of a single F. When this happens, the Fe site is completely poisoned on both sides and is no longer able to catalyze ORR.

The occurrence of single graphene layers in actual catalysts is probably quite exceptional. Those are certainly better represented by several stacks of disorganized graphene layers forming a network of connected micropores and mesopores. Therefore, we have also examined the double graphene layer case where there is a second parallel carbon layer at a distance of 3.6 Å from the upper carbon layer carrying the FeN_x sites. We found that O_2 adsorption on Fe between the two carbon layers is stable and that OOH dissociates spontaneously into O adsorbed on Fe and OH adsorbed on the opposite carbon layer. Because O_2 adsorption increases the free energy by about 1 eV (and needs an activation energy of around 1.5 eV) relative to free O_2, the catalytic process is unlikely in this case, even though the free energy of subsequent steps decreases monotonically. On the other hand, we found that F_2 can adsorb on Fe between the two carbon layers without energy expenditure, making this process more likely than for O_2. These results suggest complete poisoning of the FeN_x sites through extensive fluorination of the catalyst, in agreement with the experimental observations.

We then focused on the residual catalytic activity after fluorination by considering Fe-free N-doped carbon armchair and zigzag structures for which previous DFT calculations suggested a viable catalytic process although O_2 hardly adsorbs on these structures. For both structures, the active catalytic site is a carbon atom near a N atom. We found that these catalytic structures are not poisoned by F or F_2, thus justifying a residual ORR catalytic activity similar to that of the Fe-free catalysts observed for fluorinated Fe/N/C catalysts.

Finally, we provided an explanation for the recovery of ORR upon heating to 900 °C after fluorination. This explanation is based on the presence of radicals or small molecules released from the catalyst surface upon heat treatment. Most of the calculations presented in this work are based on free energy levels that only indicate whether a catalytic process is thermodynamically viable or not. A more thorough study would include the determination of activation energies. These calculations are very computationally demanding and will be the subject of future work.

Author Contributions: Conceptualization, M.C., J.-P.D. and F.V.; funding acquisition, S.S. and F.V.; investigation, M.C. and V.P.G.; methodology, M.C., J.-P.D. and F.V.; supervision, F.V.; writing—original draft, M.C.; writing—review and editing, M.C., J.-P.D., G.Z., V.P.G., S.S. and F.V. All authors have read and agreed to the published version of the manuscript.

Funding: This research was funded by the Natural Sciences and Engineering Research Council of Canada under grant STP 521582-18.

Data Availability Statement: Additional data are available upon request from the corresponding author. The data set generated by the simulations performed in this work has not been made publicly available due to its large quantity and diversity.

Conflicts of Interest: The authors declare no conflict of interest.

Sample Availability: No Samples are available.

Appendix A

Table A1. Lengths in Å of the bonds involving the F atoms in Figure 3a–e.

Structure/Bond	F–Fe	F–C	F–Fe–F
a	–	–	1.74, 1.98
b	1.77	1.66	–
c	1.76	1.64	–
d	–	–	1.78, 2.05
e	1.79	–	–

Table A2. Bond lengths in Å for free O_2 and for adsorbates on the structure of Figure 1j.

Structure/Bond	F–Fe	O–Fe–O	OBO
O_2	–	–	1.23
$Fe–O_2$	–	1.88, 1.88	1.35
$F–Fe–O_2$	1.80	1.92, 1.92	1.32

Table A3. F–C bond lengths in Å for sites 1 to 5 shown in the structures of Figure 2.

Structure/Position	Site 1	Site 2	Site 3	Site 4	Site 5
Armchair	1.51	1.56	1.52	1.69	1.53
Zigzag	1.44	1.49	1.53	1.67	1.50

References

1. Debe, M.K. Electrocatalyst approaches and challenges for automotive fuel cells. *Nature* **2012**, *486*, 43–51. [CrossRef] [PubMed]
2. Jaouen, F.; Herranz, J.; Lefèvre, M.; Dodelet, J.P.; Kramm, U.I.; Herrmann, I.; Bogdanoff, P.; Maruyama, J.; Nagaoka, T.; Garsuch, A.; et al. Cross-Laboratory Experimental Study of Non-Noble-Metal Electrocatalysts for the Oxygen Reduction Reaction. *ACS Appl. Mater. Interfaces* **2009**, *1*, 1623–1639. [CrossRef] [PubMed]
3. Jaouen, F.; Proietti, E.; Lefèvre, M.; Chénitz, R.; Dodelet, J.P.; Wu, G.; Chung, H.T.; Johnson, C.M.; Zelenay, P. Recent advances in non-precious metal catalysis for oxygen-reduction reaction in polymer electrolyte fuel cells. *Energy Environ. Sci.* **2011**, *4*, 114–130. [CrossRef]
4. Jaouen, F. Heat-treated transition $Metal-N_x-C_y$ electrocatalysts for the O_2 reduction reaction in acid PEM fuel cells. In *Non-Noble Metal Fuel Cell Catalysts*; Chen, Z., Dodelet, J.P., Zhang, J., Eds.; Wiley-VCH: Weinheim, Germany, 2014; Chapter 2; pp. 29–117.
5. Shao, M.; Chang, Q.; Dodelet, J.P.; Chenitz, R. Recent Advances in Electrocatalysts for Oxygen Reduction Reaction. *Chem. Rev.* **2016**, *116*, 3594–3657. [CrossRef] [PubMed]
6. Banham, D.; Ye, S.; Pei, K.; Ozaki, J.; Kishimoto, T.; Imashiro, Y. A review of the stability and durability of non-precious metal catalysts for the oxygen reduction reaction in proton exchange membrane fuel cells. *J. Power Sources* **2015**, *285*, 334–348. [CrossRef]
7. Shao, Y.; Dodelet, J.P.; Wu, G.; Zelenay, P. PGM-Free Cathode Catalysts for PEM Fuel Cells: A Mini-Review on Stability Challenges. *Adv. Mater.* **2019**, *31*, 1807615. [CrossRef]
8. Lefèvre, M.; Proietti, E.; Jaouen, F.; Dodelet, J.P. Iron-Based Catalysts with Improved Oxygen Reduction Activity in Polymer Electrolyte Fuel Cells. *Science* **2009**, *324*, 71–74. [CrossRef] [PubMed]
9. Proietti, E.; Jaouen, F.; Lefèvre, M.; Larouche, N.; Tian, J.; Herranz, J.; Dodelet, J.P. Iron-based cathode catalyst with enhanced power density in polymer electrolyte membrane fuel cells. *Nat. Commun.* **2011**, *2*, 416. [CrossRef] [PubMed]
10. Martinez, U.; Babu, S.K.; Holby, E.F.; Chung, H.T.; Yin, X.; Zelenay, P. Progress in the Development of Fe-Based PGM-Free Electrocatalysts for the Oxygen Reduction Reaction. *Adv. Mater.* **2019**, *31*, 1806545. [CrossRef]
11. Asset, T.; Atanassov, P. Iron-Nitrogen-Carbon Catalysts for Proton Exchange Membrane Fuel Cells. *Joule* **2020**, *4*, 33–44. [CrossRef]
12. Dodelet, J.P. The controversial role of the metal in Fe- or Co-based electrocatalysts for the oxygen reduction reaction in acid medium. In *Electrocatalysis in Fuel Cells, A Non- and Low- Platinum Approach*; Shao, M., Ed.; Springer-Verlag: London, UK, 2013; Chapter 10; pp. 271–338.
13. Zhang, H.; Ding, S.; Hwang, S.; Zhao, X.; Su, D.; Xu, H.; Yang, H.; Wu, G. Atomically Dispersed Iron Cathode Catalysts Derived from Binary Ligand-Based Zeolitic Imidazolate Frameworks with Enhanced Stability for PEM Fuel Cells. *J. Electrochem. Soc.* **2019**, *166*, F3116–F3122. [CrossRef]
14. Xiao, F.; Xu, G.L.; Sun, C.J.; Xu, M.; Wen, W.; Wang, Q.; Gu, M.; Zhu, S.; Li, Y.; Wei, Z.; et al. Nitrogen-coordinated single iron atom catalysts derived from metal organic frameworks for oxygen reduction reaction. *Nano Energy* **2019**, *61*, 60–68. [CrossRef]
15. Kattel, S.; Wang, G. Reaction Pathway for Oxygen Reduction on FeN_4 Embedded Graphene. *J. Phys. Chem. Lett.* **2014**, *5*, 452–456. [CrossRef]
16. Holby, E.F.; Wu, G.; Zelenay, P.; Taylor, C.D. Structure of $Fe-N_x-C$ Defects in Oxygen Reduction Reaction Catalysts from First-Principles Modeling. *J. Phys. Chem. C* **2014**, *118*, 14388–14393. [CrossRef]
17. Holby, E.F.; Zelenay, P. Linking structure to function: The search for active sites in non-platinum group metal oxygen reduction reaction catalysts. *Nano Energy* **2016**, *29*, 54–64. [CrossRef]
18. Matanovic, I.; Artyushkova, K.; Atanassov, P. Understanding PGM-free catalysts by linking density functional theory calculations and structural analysis: Perspectives and challenges. *Curr. Opin. Electrochem.* **2018**, *9*, 137–144. [CrossRef]
19. Ji, Y.; Dong, H.; Liu, C.; Li, Y. The progress of metal-free catalysts for the oxygen reduction reaction based on theoretical simulations. *J. Mater. Chem. A* **2018**, *6*, 13489–13508. [CrossRef]
20. Chenitz, R.; Kramm, U.I.; Lefèvre, M.; Glibin, V.P.; Zhang, G.; Sun, S.; Dodelet, J.P. A specific demetalation of $Fe-N_4$ catalytic sites in the micropores of NC_Ar+NH_3 is at the origin of the initial activity loss of the highly active Fe/N/C catalyst used for the reduction of oxygen in PEM fuel cells. *Energy Environ. Sci.* **2018**, *11*, 365–382. [CrossRef]

21. Singh, D.; Tian, J.; Mamtani, K.; King, J.; Miller, J.T.; Ozkan, U.S. A comparison of N-containing carbon nanostructures (CNx) and N-coordinated iron-carbon catalysts (FeNC) for the oxygen reduction reaction in acidic media. *J. Catal.* **2014**, *317*, 30–43. [CrossRef]

22. Mamtani, K.; Ozkan, U.S. Heteroatom-Doped Carbon Nanostructures as Oxygen Reduction Reaction Catalysts in Acidic Media: An Overview. *Catal. Lett.* **2015**, *145*, 436–450. [CrossRef]

23. Mamtani, K.; Singh, D.; Tian, J.; Miller, J.M.; Miller, J.T.; Co, A.C.; Ozkan, U.S. Evolution of N-Coordinated Iron-Carbon (FeNC) Catalysts and Their Oxygen Reduction (ORR) Performance in Acidic Media at Various Stages of Catalyst Synthesis: An Attempt at Benchmarking. *Catal. Lett.* **2016**, *146*, 1749–1770. [CrossRef]

24. Saputro, A.G.; Kasai, H. Density Functional Theory Study on the Interaction of O_2 and H_2O_2 Molecules with the Active Sites of Cobalt-Polypyrrole Catalyst. *J. Phys. Soc. Jpn.* **2014**, *83*, 024707. [CrossRef]

25. Martinez, U.; Babu, S.K.; Holby, E.F.; Zelenay, P. Durability challenges and perspective in the development of PGM-free electrocatalysts for the oxygen reduction reaction. *Curr. Opin. Electrochem.* **2018**, *9*, 224–232. [CrossRef]

26. Yin, X.; Zelenay, P. Kinetic models for the degradation mechanisms of PGM-free ORR catalysts. *ECS Trans.* **2018**, *85*, 1239–1250. [CrossRef]

27. Wang, X.X.; Prabhakaran, V.; He, Y.; Shao, Y.; Wu, G. Iron-Free Cathode Catalysts for Proton-Exchange-Membrane Fuel Cells: Cobalt Catalysts and the Peroxide Mitigation Approach. *Adv. Mater.* **2019**, *31*, 1805126. [CrossRef]

28. Zhang, J.; Zhou, Z.; Wang, F.; Li, Y.; Jing, Y. Two-dimensional metal hexahydroxybenzene frameworks as promising electrocatalysts for an oxygen reduction reaction. *ACS Sustain. Chem. Eng.* **2020**, *8*, 7472–7479. [CrossRef]

29. Li, T.; Li, M.; Zhu, X.; Zhang, J.; Jing, Y. Conductive two-dimensional $M_3(C_6S_3O_3)_2$ monolayers as effective electrocatalysts for the oxygen reduction reaction. *J. Mater. Chem. A* **2021**, *9*, 24887–24894. [CrossRef]

30. Berthon-Fabry, S.; Dubau, L.; Ahmad, Y.; Guerin, K.; Chatenet, M. First Insight into Fluorinated Pt/Carbon Aerogels as More Corrosion-Resistant Electrocatalysts for Proton Exchange Membrane Fuel Cell Cathodes. *Electrocatalysis* **2015**, *6*, 521–533. [CrossRef]

31. Asset, T.; Chattot, R.; Maillard, F.; Dubeau, L.; Ahmad, Y.; Batisse, N.; Dubois, M.; Guérin, K.; Labbé, F.; Metkemeijer, R.; et al. Activity and Durability of Platinum-Based Electrocatalysts Supported on Bare or Fluorinated Nanostructured Carbon Substrates. *J. Electrochem. Soc.* **2018**, *165*, F3346–F3358. [CrossRef]

32. Wang, Y.C.; Zhu, P.F.; Yang, H.; Huang, L.; Wu, Q.H.; Rauf, M.; Zhang, J.Y.; Dong, J.; Wang, K.; Zhou, Z.Y.; et al. Surface Fluorination to Boost the Stability of the Fe/N/C Cathode in Proton Exchange Membrane Fuel Cells. *ChemElectroChem* **2018**, *5*, 1914–1921. [CrossRef]

33. Zhang, G.; Yang, X.; Dubois, M.; Herraiz, M.; Chenitz, R.; Lefèvre, M.; Cherif, M.; Vidal, F.; Glibin, V.P.; Sun, S.; et al. Non-PGM electrocatalysts for PEM fuel cells: Effect of fluorination on the activity and stability of a highly active NC_Ar+NH$_3$ catalyst. *Energy Environ. Sci.* **2019**, *12*, 3015–3037. [CrossRef]

34. Li, Y.; Liu, X.; Zheng, L.; Shang, J.; Wan, X.; Hu, R.; Guo, X.; Hong, S.; Shui, J. Preparation of Fe-N-C catalysts with FeN$_x$ (x = 1, 3, 4) active sites and comparison of their activities for the oxygen reduction reaction and performances in proton exchange membrane fuel cells. *J. Mater. Chem. A* **2019**, *7*, 26147–26153. [CrossRef]

35. Saputro, G.; Kasai, A.; Asazawa, H.; Kishi, K.; Tanaka, H. Comparative study on the catalytic activity of the TM-N2 active sites (TM = Mn, Fe, Co, Ni) in the oxygen reduction reaction: Density functional theory study. *J. Phys. Soc. Jpn.* **2013**, *82*, 114704. [CrossRef]

36. Xia, D.; Yang, X.; Xie, L.; Wei, Y.; Jiang, W.; Dou, M.; Li, X.; Li, J.; Gan, L.; Kang, F. Direct Growth of Carbon Nanotubes Doped with Single Atomic Fe-N4 Active Sites and Neighboring Graphitic Nitrogen for Efficient and Stable Oxygen Reduction Electrocatalysis. Adv. *Funct. Mater.* **2019**, *29*, 1906174. [CrossRef]

37. Anderson, A.B.; Holby, E.F. Pathways for O_2 Electroreduction over Substitutional FeN$_4$, HOFeN$_4$, and OFeN$_4$ in Graphene Bulk Sites: Critical Evaluation of Overpotential Predictions Using LGER and CHE Models. *J. Phys. Chem. C* **2019**, *123*, 18398–18409. [CrossRef]

38. Yan, M.; Dai, Z.; Chen, S.; Dong, L.; Zhang, X.L.; Xu, Y.; Sun, C. Single-Iron Supported on Defective Graphene as Efficient Catalysts for Oxygen Reduction Reaction. *J. Phys. Chem. C* **2020**, *124*, 13283–13290. [CrossRef]

39. Liu, K.; Qiao, Z.; Hwang, S.; Liu, Z.; Zhang, H.; Su, D.; Xu, H.; Wu, G.; Wang, G. Mn- and N-doped carbon as promising catalysts for oxygen reduction reaction: Theoretical prediction and experimental validation. *Appl. Catal. B-Environ.* **2019**, *243*, 195–203. [CrossRef]

40. Li, J.; Chen, M.; Cullen, D.A.; Hwang, S.; Wang, M.; Li, B.; Liu, K.; Karakalos, S.; Lucero, M.; Zhang, H.; et al. Atomically dispersed manganese catalysts for oxygen reduction in proton-exchange membrane fuel cells. *Nat. Catal.* **2018**, *1*, 935–945. [CrossRef]

41. Song, J.; Ren, Y.; Li, J.; Huang, X.; Cheng, F.; Tang, Y.; Wang, H. Core-shell Co/CoNx@C nanoparticles enfolded by Co-N doped carbon nanosheets as a highly efficient electrocatalyst for oxygen reduction reaction. *Carbon* **2018**, *138*, 300–308. [CrossRef]

42. Kattel, S.; Atanassov, P.; Kiefer, B. Catalytic activity of Co-Nx/C electrocatalysts for oxygen reduction reaction: A density functional theory study. *Phys. Chem. Chem. Phys.* **2013**, *15*, 148–153. [CrossRef] [PubMed]

43. Nørskov, J.K.; Rossmeisl, J.; Logadottir, A.; Lindqvist, L.; Kitchin, J.R.; Bligaard, T.; Jonsson, H. Origin of the overpotential for oxygen reduction at a fuel-cell cathode. *J. Phys. Chem. B* **2004**, *108*, 17886–17892. [CrossRef]

44. Zhang, P.; Xiao, B.B.; Hou, X.L.; Zhu, Y.F.; Jiang, Q. Layered SiC Sheets: A Potential Catalyst for Oxygen Reduction Reaction. *Sci. Rep.* **2015**, *4*, 3821. [CrossRef] [PubMed]

45. Li, M.; Zhang, L.; Xu, Q.; Niu, J.; Xia, Z. N-doped graphene as catalysts for oxygen reduction and oxygen evolution reactions: Theoretical considerations. *J. Catal.* **2014**, *314*, 66–72. [CrossRef]

46. Zhang, J.; Wang, Z.; Zhu, Z. A density functional theory study on oxygen reduction reaction on nitrogen-doped graphene. *J. Mol. Model.* **2013**, *19*, 5515–5521. [CrossRef] [PubMed]

47. Ikeda, T.; Hou, Z.; Chai, G.L.; Terakura, K. Possible Oxygen Reduction Reactions for Graphene Edges from First Principle. *J. Phys. Chem. C* **2014**, *118*, 17616–17625. [CrossRef]

48. Zhang, H.; Zhao, J.; Cai, Q. Pyridine derivative/graphene nanoribbon composites as molecularly tunable heterogeneous electrocatalysts for the oxygen reduction reaction. *Phys. Chem. Chem. Phys.* **2016**, *18*, 5040–5047. [CrossRef]

49. Studt, F. The Oxygen Reduction Reaction on Nitrogen-Doped Graphene. *Catal. Lett.* **2013**, *143*, 58–60. [CrossRef]

50. Duan, Z.; Henkelman, G. Identification of Active Sites of Pure and Nitrogen-Doped Carbon Materials for Oxygen Reduction Reaction Using Constant-Potential Calculations. *J. Phys Chem. C* **2020**, *124*, 12016–12023. [CrossRef]

51. Paul, R.; Zhu, L.; Chen, H.; Qu, J.; Dai, L. Recent Advances in Carbon-Based Metal-Free Electrocatalysts. *Adv. Mater.* **2019**, *31*, 1806403. [CrossRef] [PubMed]

52. Glibin, V.P.; Cherif, M.; Vidal, F.; Dodelet, J.P.; Zhang, G.; Sun, S. Non-PGM Electrocatalysts for PEM Fuel Cells: Thermodynamic Stability and DFT Evaluation of Fluorinated FeN4-Based ORR Catalysts. *J. Electrochem. Soc.* **2019**, *166*, F3277–F3286. [CrossRef]

53. Svane, K.L.; Reda, M.; Vegge, T.; Hansen, H.A. Improving the activity of M-N$_4$ catalysts for the oxygen reduction reaction by electrolyte adsorption. *ChemSusChem* **2019**, *12*, 5133–5141. [CrossRef] [PubMed]

54. Rebarchik, M.; Bhandari, S.; Kropp, T.; Mavrikakis, M. How noninnocent spectator species improve the oxygen reduction activity of single-atom catalysts: Microkinetic models from first-principles calculations. *ACS Catal.* **2020**, *10*, 9129–9135. [CrossRef]

55. Yang, Y.; Li, K.; Meng, Y.; Wang, Y.; Wu, Z. A density functional study on the oxygen reduction reaction mechanism on FeN$_2$-doped graphene. *New J. Chem.* **2018**, *42*, 6873–6879. [CrossRef]

56. Szakacs, C.E.; Lefèvre, M.; Kramm, U.I.; Dodelet, J.P.; Vidal, F. A density functional theory study of catalytic sites for oxygen reduction in Fe/N/C catalysts used in H$_2$/O$_2$ fuel cells. *Phys. Chem. Chem. Phys.* **2014**, *16*, 13654–13661. [CrossRef] [PubMed]

57. Hesse, W.M.; Herd, C.R. Microstructure, Morphology and General Physical Properties. In *Carbon Black*, 2nd ed.; Donnet, J.B., Bansal, R.C., Wang, M.J., Eds.; Marcel Dekker: New York, NY, USA, 1993; Chapter 3; pp. 89–173.

58. Clementi, E.; Raimondi, D.L.; Reinhardt, W.P. Atomic Screening Constants from SCF Functions. II. Atoms with 37 to 86 Electrons. *J. Chem. Phys.* **1967**, *47*, 1300–1307. [CrossRef]

59. Zhang, G.; Chenitz, G.; Lefèvre, M.; Sun, S.; Dodelet, J.P. Is iron involved in the lack of stability of Fe/N/C electrocatalysts used to reduce oxygen at the cathode of PEM fuel cells? *Nano Energy* **2016**, *29*, 111–125. [CrossRef]

60. Dubois, M.; Guérin, K.; Ahmad, Y.; Batisse, N.; Mar, M.; Frezet, L.; Hourani, W.; Bubendorff, J.L.; Parmentier, J.; Hajar-Garreau, S.; et al. Thermal exfoliation of fluorinated graphite. *Carbon* **2014**, *77*, 688–704. [CrossRef]

61. Kresse, G.; Hafner, J. Ab-initio molecular dynamics for liquid metals. *Phys. Rev. B* **1993**, *47*, 558–561. [CrossRef] [PubMed]

62. Kresse, G.; Hafner, J. Ab-initio molecular dynamics simulation of the liquid metal amorphous semiconductor transition in germanium. *Phys. Rev. B* **1994**, *49*, 14251–14269. [CrossRef]

63. Kresse, G.; Furthmüller, J. Efficiency of ab-initio total energy calculations for metals and semiconductors using a plane-wave basis set. *Comput. Mat. Sci.* **1996**, *6*, 15–50. [CrossRef]

64. Kresse, G.; Furthmüller, J. Efficient iterative schemes for ab initio total-energy calculations using a plane-wave basis set. *Phys. Rev. B* **1996**, *54*, 11169–11186. [CrossRef] [PubMed]

65. Perdew, J.P.; Burke, K.; Ernzerhof, M. Generalized gradient approximation made simple. *Phys. Rev. Lett.* **1996**, *77*, 3865–3868. [CrossRef] [PubMed]

66. Lyalin, A.; Nakayama, A.; Uosaki, K.; Taketsugu, T. Theoretical predictions for hexagonal BN based nanomaterials as electrocatalysts for the oxygen reduction reaction. *Phys. Chem. Chem. Phys.* **2013**, *15*, 2809–2820. [CrossRef] [PubMed]

Article

A Nanosheet-Assembled SnO$_2$-Integrated Anode

Xiaoli Wang [1], **Xinyu Zhao** [2,*] **and Yin Wang** [2,*]

[1] Liaoning Key Laboratory of Chemical Additive Synthesis and Separation, Department of Chemistry and Environment Engineering, Yingkou Institute of Technology, Yingkou 115014, China; wangxldq@sina.com
[2] Inner Mongolia Key Laboratory of Carbon Nanomaterials, College of Chemistry and Materials Science, Inner Mongolia Minzu University, Tongliao 028000, China
* Correspondence: xyzhao@imun.edu.cn (X.Z.); ywang@imun.edu.cn (Y.W.)

Abstract: There is an ever-increasing trend toward bendable and high-energy-density electrochemical storage devices with high strength to fulfil the rapid development of flexible electronics, but they remain a great challenge to be realised by the traditional slurry-casting fabrication processes. To overcome these issues, herein, a facile strategy was proposed to design integrating an electrode with flexible, high capacity, and high tensile strength nanosheets with interconnected copper micro-fibre as a collector, loaded with a novel hierarchical SnO$_2$ nanoarchitecture, which were assembled into core–shell architecture, with a 1D micro-fibre core and 2D nanosheets shell. When applied as anode materials for LIBs, the resultant novel electrode delivers a large reversible specific capacity of 637.2 mAh g^{-1} at a high rate of 1C. Such superior capacity may benefit from rational design based on structural engineering to boost synergistic effects of the integrated electrode. The outer shell with the ultrathin 2D nanoarchitecture blocks can provide favourable Li$^+$ lateral intercalation lengths and more beneficial transport routes for electrolyte ions, with sufficient void space among the nanosheets to buffer the volume expansion. Furthermore, the interconnected 1D micro-fibre core with outstanding metallic conductivity can offer an efficient electron transport pathway along axial orientation to shorten electron transport. More importantly, the metal's remarkable flexibility and high tensile strength provide the hybrid integrated electrode with strong bending and stretchability relative to sintered carbon or graphene hosts. The presented strategy demonstrates that this rational nanoarchitecture design based on integrated engineering is an effective route to maintain the structural stability of electrodes in flexible LIBs.

Keywords: anode; flexible electronics; nanosheets; SnO$_2$

Citation: Wang, X.; Zhao, X.; Wang, Y. A Nanosheet-Assembled SnO$_2$-Integrated Anode. *Molecules* **2021**, *26*, 6108. https://doi.org/10.3390/molecules26206108

Academic Editors: Gregorio F. Ortiz and Munkhbayar Batmunkh

Received: 1 August 2021
Accepted: 7 October 2021
Published: 10 October 2021

Publisher's Note: MDPI stays neutral with regard to jurisdictional claims in published maps and institutional affiliations.

1. Introduction

Nowadays, an urgent and key task for energy conversion storage systems, in particular lithium-ion batteries (LIBs), is to develop advanced electrode materials with mechanical durability and superior Li-storage performance for booming flexible energy storage applications in foldable smartphones, wearable electronic systems, and implantable device [1–5]. However, the design and fabrication of such a high-performance flexible electrode is still a major challenge via a facile method because of the lack of optimal materials with the feature of high special capacity and robust mechanical flexibility in electrochemical environments. The traditional slurry-coating fabrication technology is not suitable for flexible LIBs because the active materials often suffer from exfoliation or cracking in the process of frequent bending. Furthermore, these additional binders in slurry would hinder electronic transport and reduce the specific energy density of the battery, and conductive carbon black should be added into the slurry to improve electrical performance.

In order to keep pace with the development of flexible energy storage systems and solve these problems of excessive consumption of adhesives and carbon black (~30 wt%), one of the most effective strategies is to construct an integrated electrode instead of slurry-coating fabrication technology [6–11]. In this regard, searching for an appropriate flexible current collector and high-energy-density active material is crucial to achieving these goals. To date, noteworthy progress in novel flexible electrodes has been achieved through chemical synthesis routes, including hydrothermal [12], sol–gel techniques [13], and CVD [14]. Nevertheless, most reported flexible energy storage devices are performed by using bio-driven carbon or free-standing film, which are not strong enough to withstand frequent mechanical deformations during practical use. Hence, an integrated electrode with robust tensile strength and enhanced electrochemical performance is still plagued and needs to be further improved.

Moreover, group IV element Sn in the form of metal oxide (SnO_2) has particularly attracted extensive attention as a promising anode material to replace conventional graphitic carbon in current LIBs because of its uniqueness in terms of low cost, safe working potential, high theoretical capacity, and environmental friendliness [15–19]. Nevertheless, the simple structure, relatively low intrinsic conductivity, and vast structural variation during the reversible insertion/deinsertion processes of the bulk SnO_2 powder keep it from achieving its full capacitance potential. One of the effective strategies to mitigate these problems is to develop structural engineering, including morphology control and hybrid construction, which could alleviate the mechanical stress induced by large volume change and prevent aggregation of the active domains [20–25]. However, these materials still need to be mixed with a binder and carbon black and pressed onto metal substrates or, alternatively, by being deposited onto a conductive substrate before they are assembled into batteries, which makes them less flexible and have a low energy density. Integrated electrodes, in which electrochemically active nanostructures are conformably coated on conductive collectors, have been demonstrated with ultrafast power rate and long lifespan. The successful integration of the sturdy conductive matrix support with elegant nanostructures improves the electrode performance and endows it with robust mechanical flexibility. All these merits render this type of electrode very attractive for flexible power sources.

To combine the aforementioned merits and promote the development of flexible LIBs, herein, we devised a novel route to fabricate SnO_2-integrated electrode assembled by core–shell architecture, with a 1D micro-fibre core and 2D nanosheets shell, targeting high capacity and tensile strength LIBs. The copper micro-fibre clothes function as a superior conductive pathway facilitating fast electrons transfer along the axial orientation but also provide a high mechanical substrate assisting independent growth of SnO_2 nanosheets. The nanosheets shells are vertically distributed on copper micro-fibre, forming core–shell structure, which can provide a large electrode/electrolyte interfacial contact area. As expected, when measured as anode materials of LIBs, we obtained a reversible lithium storage capacity of 637.2 mAh g^{-1} at a current density of 1C. The presented synthetic strategy is effective and with low cost, which provides a novel route to design advanced electrode materials for flexible LIBs.

2. Materials and Methods

2.1. Sample Synthesis

Prior to the synthesis, a piece of copper micro-fibre (CMF) textile (approximately 5×5 cm^2) (Liaoning Copper Group, Liaoyang, China) was treated by ultrasonication with 1 M hydrochloric acid solution (Aladdin, Shanghai, China), in order to remove the CuO layer. In a typical hydrothermal synthesis of the SnO_2 nanosheets array, 12 mmol $SnCl_2 \bullet 2H_2O$ and 24 mmol NH_4F were first mixed in 70 mL deionised water under magnetic stirring (Aladdin, China). The clear and transparent solution was continuously stirred for 0.5 h in the air. Then, the mixed solution was transferred into a 100 mL Teflon-lined stainless steel autoclave. Afterwards, the CMF substrate was immersed in the solution. The sealed autoclave was heated to 180 °C for 24 h. Subsequently, the grey cloth was cleaned repeatedly with deionised water and ethanol under ultrasonic treatment and then dried under N_2 gas flow. To investigate the formation of the nanosheet SnO_2, a series of parallel experiments were carried out by adjusting the molar ratio of fluoride/Sn.

2.2. Materials Characterisation

The crystallographic information of the as-prepared materials was recorded with powder X-ray diffractometry (XRD, Bruker D8 Advance, Bremen, Germany). The morphological features and microstructure of the sample were observed by using a field-emission scanning electron microscopy (FE-SEM, S-4800, Hitachi, Tokyo, Japan) and transmission electron microscopy (TEM, JEOL JEM-2100F, Japan). X-ray photoelectron spectroscopic (XPS, ESCALAB 250, Thermo Scientific, Waltham, MA, USA) was used to characterise the surface composition of the sample.

2.3. Electrochemical Measurements

Electrochemical properties were measured using coin cells (CR2025), which were assembled in an argon-filled glovebox (Mbraun, Unilab, Germany). The copper micro-fibre-textile-supported ultrathin SnO_2 nanosheets were used directly as the working electrode for the subsequent electrochemical tests without binders and conductivity carbon black. The coated cloth was cut into disk electrodes (12 mm in diameter). To fabricate the SnO_2 power working electrode, the active materials, super-P, and PVDF with a mass ratio of 80:10:10 were ground in NMP solvent to form a homogeneous slurry, which was then coated onto the Cu foil by the doctor blade method and dried by heating in a vacuum oven (Yiheng, Shanghai, China). The lithium foil was used as the cathode electrode. The commercial electrolyte in the present measurements was a mixture of $LiPF_6$ in ethylene carbonate, dimethyl carbonate, and diethyl carbonate (EC-DEC-EMC, 1:1:1 in v/v). Galvanostatic cycling performances of the as-prepared coin cell were operated at room temperature on a multi-channel battery testing system (Land CT2001A, Wuhan, China) with a cut-off voltage of 1.2–0.01 V versus Li/Li^+. Cyclic voltammetry (CV) curves were carried out by applying a CHI-760E electrochemical workstation at a scanning rate of 0.1 mV s^{-1}.

3. Results and Discussion

The flexible integrated electrode samples were fabricated via a facile in situ hydrothermal depositions of active material on the CMF without any post-treatment and use of conventional carbon black additive and binder. The overall fabrication process of the SnO_2-integrated electrode is shown schematically in Figure 1. As detailed in the Experimental Section, the acid-treated copper micro-fibre textile was immersed in a mixed solution of stannous chloride dihydrate and ammonium fluoride, followed by a hydrothermal reaction at 180 °C for 24 h, yielding the large-size free-standing SnO_2 nanosheets on the CMF. The SnO_2-integrated electrode materials display superior flexibility and robust mechanical tensile, which can be cut directly into electrode pieces with a diameter of 12 mm.

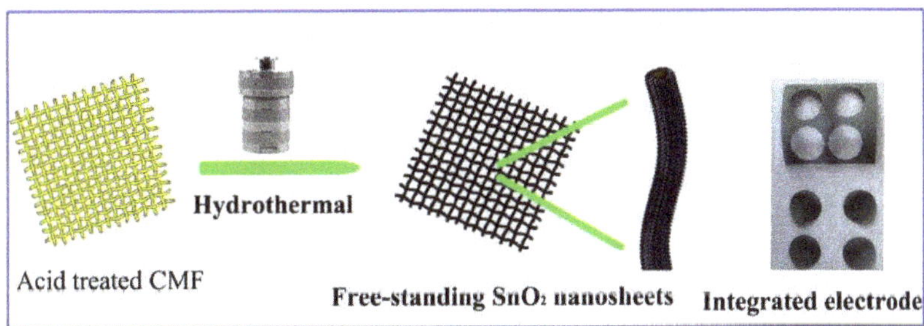

Figure 1. A schematic diagram for the fabrication procedure of SnO_2 nanosheets on the CMF.

Firstly, the powder X-ray diffraction (XRD) technique was provided to determine the composition characterisation of the as-prepared SnO_2 samples. Figure 2a displays the crystallinity and phase purity of the SnO_2 nanosheets assembled integration materials. The main three peaks at around 43.3, 50.4, and 74.1° correspond to the metal copper substrate. Removing the Cu element signals from the substrates, all the diffraction peaks can be well indexed to the tetragonal rutile structure of SnO_2 (JCPDS Card no.41-1445). No impurities, such as SnO or Sn, were detected, indicating the formation of pure SnO_2 nanosheets. In addition, a piece of SnO_2/CMF-integrated composite sample was put into a concentrated nitric acid solution to remove the Cu substrates, and the resulting powers were purified with deionised water and ethanol for subsequent testing. As shown in Figure 2b, the XRD pattern indicated that the as-prepared powers after concentrated nitric acid treatment were in a pure tetragonal rutile crystalline phase. Supplementary Figure S1 presents the XRD pattern of the SnO_2 nanoflower assembled nanosheets synthesised by hydrothermal method without CMF, which is in accordance with the above-mentioned sample.

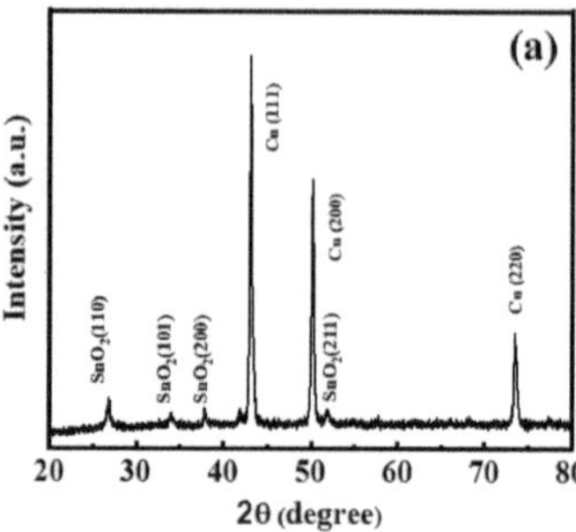
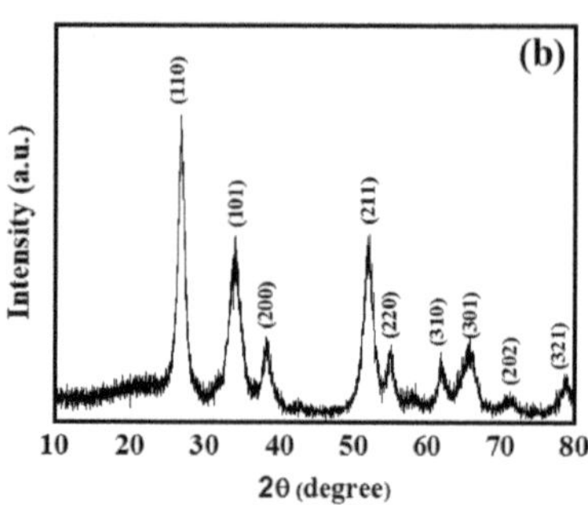

Figure 2. XRD pattern of SnO_2: (**a**) nanosheets on the CMF; (**b**) nanosheets after HNO_3 treatment.

To further verify the near-surface chemical composition and the oxidation state of element Sn in the as-prepared SnO_2 products, XPS measurement was then performed, as shown in Figure 3. A survey XPS spectrum of SnO_2 nanosheet is clearly observed in Figure 3a, indicating the existence of Sn and O elements. For the high resolution of element Sn, two strong peaks centred at 486.8 and 495.2 eV in the XPS spectrum could match with Sn $3d_{5/2}$ and $3d_{3/2}$ states, respectively. This clearly indicated that tin was in the Sn (IV) state in the nanosheets sample, which is in good accordance with previously reported SnO_2 [26].

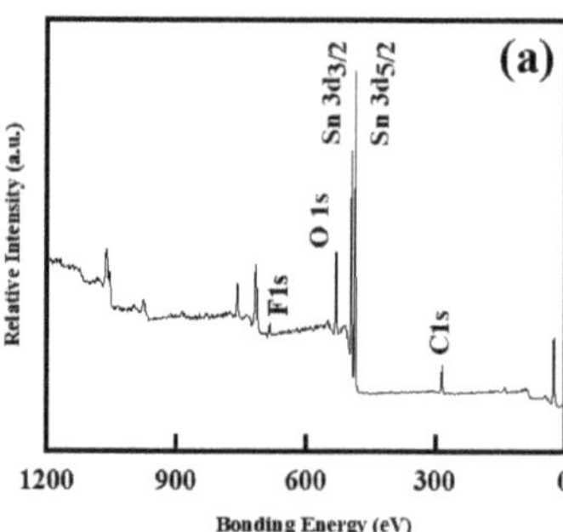
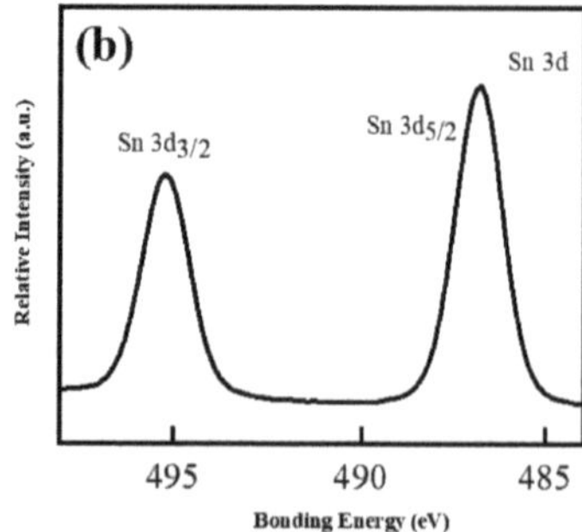

Figure 3. XPS spectrum of SnO_2 nanosheets: (**a**) survey XPS pattern; (**b**) high-resolution XPS of Sn 3d.

FE-SEM and TEM images with different magnifications could provide information about the surface morphology and crystallographic properties of the as-received products. Figure 4a shows a typical SEM image of the uncovered SnO_2 nanosheets micro-fibre cloth composed of perpendicular and smooth copper micro-fibre, which was interconnected into a textile structure (20×20 μm^2 squared pore) and served as the backbone for the growth of SnO_2 nanosheets. As shown from Figure 4b, the hydrothermal treatment resulted in a significant morphology change of the copper micro-fibre cloth from a relatively smooth surface to a very rough surface, which indicated that the conductive substrate had been covered with the numerous SnO_2 nanosheets and formed a CMF@SnO_2 hybrid material. Based on recorded top-view high magnification SEM images (Figure 4c–e), the outer sheath in the presented 1D architecture consisted of a uniform sheet structure with a smooth surface. Apparent open space between adjacent SnO_2 nanosheets presented sponge-like porous architecture from the magnified SEM images (Figure 4d,e), which provided a large interfacial area for electrolyte ion diffusion and ensured a short solid-state diffusion length for fast Li-ion insertion/extraction. The direct growth of SnO_2 nanosheets on the copper micro-fibre network collector enabled good contact and strong binding between SnO_2 and copper micro-fibre without any binder and carbon black. Due to the high surface energy effect of nanostructured materials, a large number of nanosheets were attached to the outer of the integrated electrode (Supplementary Figure S2), which maybe bring more negative effects on the subsequent electrochemical testing. Hence, the SnO_2-integrated electrode was purified under ultrasonic treatment. As a result, SnO_2 nanosheets in situ grown on the CMF were not peeled off even after repeated bending or ultrasonic treatment due to the present interconnected nanoarchitecture. Figure 4f shows the photograph of a copper textile coated with a layer of grey SnO_2 nanosheets. Notably, similar to micro-fibre cloth, the SnO_2-integrated electrode with the nanosheets coating could be easily bent without damage to the nanosheets, making them interesting for flexible batteries.

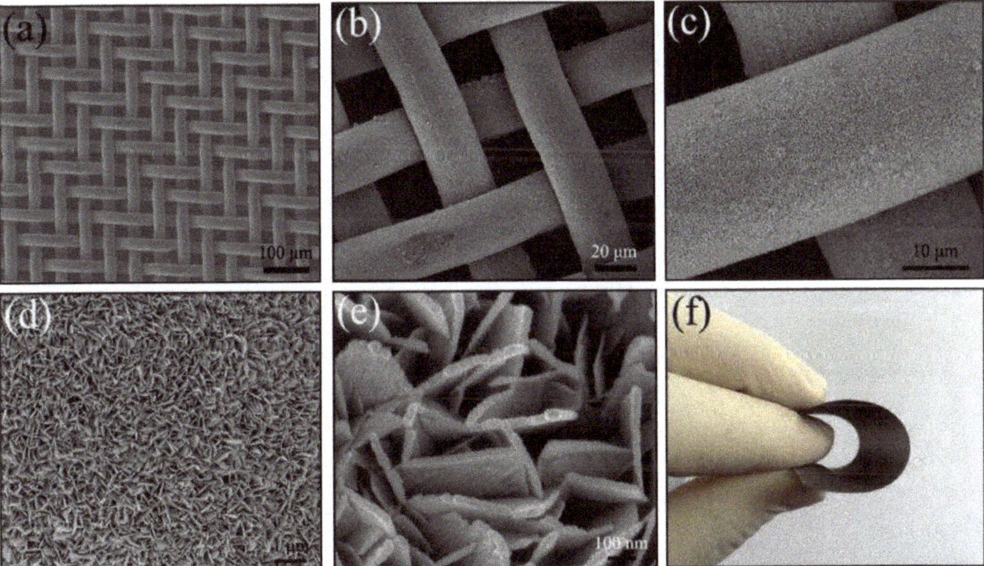

Figure 4. SEM images of (**a**) pristine copper microfibre cloth, where sub-millimetre pores can be clearly observed; (**b**) and (**c**) low magnification images of the copper textile surfaces after SnO_2 nanosheets layers are grown; (**d**) and (**e**) high-magnification SnO_2 nanosheets; (**f**) shows a photo of a piece of folded copper textile coated SnO_2 nanosheets.

Further microstructure information about SnO_2 nanosheets building blocks was obtained from transmission electron microscopy (TEM) (Figure 5). As shown in Figure 5a–d, the nanosheets structure could be clearly seen at low magnification. TEM further revealed that the hierarchical SnO_2 nanoarchitecture was built up of highly porous intercon nected nanosheets. The curled geometrical morphology further showed that the presented SnO_2 electrode materials exhibited significantly improved microstructural flexible performance through nanostructured engineering. The average lateral size and thickness of the nanosheets were found to be approximately 300 and 20 nm, respectively. The morphology and size of the sample obtained from TEM were in good agreement with those observed in the SEM images. Furthermore, Figure 5d shows the HR-TEM image and measured lattice fringes as 3.35 Å, which is consistent with the (111) interplanar distance of the SnO_2 phase. The inset of Figure 5d is the SAED pattern of a piece of SnO_2 nanosheets film, which clearly demonstrates the polycrystalline nature of the nanosheets films.

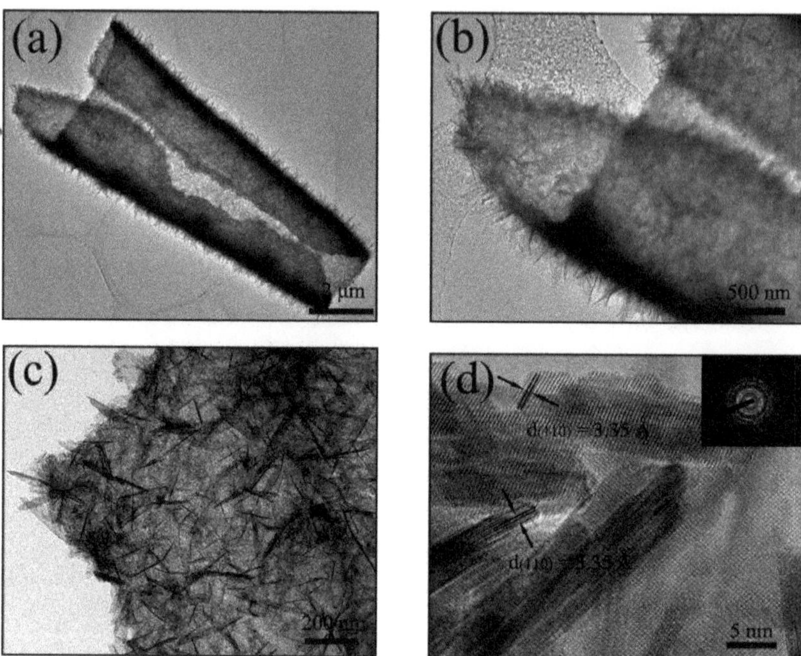

Figure 5. (**a–c**) Typical TEM images of the nanosheets SnO_2; (**d**) HRTEM images of the nanosheets SnO_2. The inset of d shows that these nanosheets are polycrystalline.

To gain insight into the effect of ammonium fluorides (NH_4F) on the morphology of SnO_2 nanostructure, fluoride-dependent experiments were carried out. As shown in Supplementary Figure S3a, SnO_2 nanoparticles with an irregular shape and size of several nanometres were obtained in the absence of NH_4F, which confirmed that the NH_4F was key to the formation of the 2D SnO_2 nanoarchitecture. When the F/Sn molar ratio was 1, hollow-nanosphere-shaped samples with coarse surfaces could be detected (Supplementary Figure S3b). Once the F/Sn molar ratio was increased to 2, the well-defined 3D flower-like SnO_2 hollow nanoarchitecture (or nanoflower) constructed by many 2D nanosheets were generated, and their detailed characteristics are described in Supplementary Figure S3c,d. However, the SnO_2 products could not be found in the Teflon-lined stainless-steel autoclave if the molar ratio was increased to 4.

To prove the efficacy of the as-synthesis unique CMF@SnO_2 nanosheets architecture as a potential anode material suitable for practical application, we next performed an electrochemical evaluation of this integration electrode without adding any binders (PVDF) and conductive black additives (CB) in half-cells. It is well known that the electrochemical reaction process in the anode can be divided into two steps, as follows:

$$SnO_2 + 4Li^+ + 4e^- \rightarrow Sn + 2Li_2O \tag{1}$$

$$Sn + xLi^+ + xe^- \leftrightarrow Li_xSn \ (0 \leq x \leq 4.4) \tag{2}$$

$$Li^+ + e^- + electrolyte \rightarrow SEI\ layer \tag{3}$$

To gain better insight into the electrochemical performance of the novel integration anode, cyclic voltammetry (CV) characteristics of the initial four cyclic were first performed between 0.01 and 3 V, at a scan rate of 0.1 mV s^{-1}. As can be seen from Figure 6a, there was a substantial difference between the first and the subsequent cycles. In the first cathodic process, there were significant reduction peaks at 0.8 to 1.0 V, which is related to the conversion of tin dioxide to metallic tin and lithium oxide, and the formation of SEI layer on the electrode surface, as illustrated in Equations (1) and (3) [23,24]. Note that there was a strong reduction peak located at around 0.05 V, which is assigned to the formation of the Li$_x$Sn alloys process (Equation (2)). In the subsequent anodic scan, a strong peak at 0.6 V corresponded to the phase transition from Li$_x$Sn alloy to metallic Sn, which increased in intensity as the cycle number was increased. This phenomenon could be explained by the activation of the reversible reaction that occurred in the electrode materials [27]. Beyond that, oxidation peaks around 1.3 and 1.9 V were also observed for the SnO$_2$-integrated anode materials [23,24,28]. This observation might suggest partial reversibility of the reaction by Equation (1), which is also observed for the nano-sized SnO$_2$ and SnO$_2$/carbon composites [29,30]. In the following process of CV testing, the peak potentials of anodic and cathodic curves were almost coincident, and the peak intensity changed very slightly, indicating robust cycling stability of SnO$_2$-integrated anode.

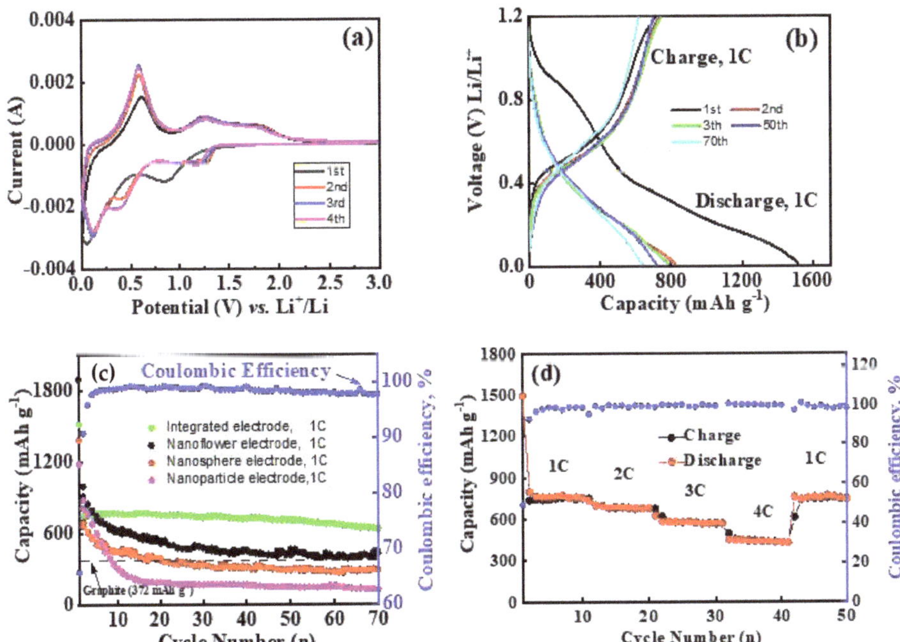

Figure 6. (**a**) Cyclic voltammograms (CVs) of nanosheet-assembled SnO$_2$-integrated electrode; (**b**) discharge–charge voltage profile of nanosheet-assembled SnO$_2$-integrated electrode between 0.01 and 1.2 V at a rate of 1 C; (**c**) cycling performance and coulombic efficiency (CE) of nanostructured SnO$_2$ electrode; (**d**) cycling performance at different rates (1–4 C) of nanosheet-assembled SnO$_2$-integrated electrode.

Furthermore, the galvanostatic charge and discharge measurements of the nanosheet-assembled SnO$_2$-integrated electrode were carried out at a high rate of 1 C for 1st, 2nd, 3rd, 50th, and 70th cycles in order to study Li-storage performance. It can be clearly seen that the nanosheet-assembled SnO$_2$-integrated electrode showed a similar curve to those of SnO$_2$ anode materials and delivered a large initial discharge and charge capacity of 1518.1 and 818.6 mAh g^{-1}, respectively. The capacity loss, compared with the first cycle, may be mainly attributed to irreversible side reactions, such as the trapping of some lithium in the lattice, formation of solid–electrolyte interface (SEI) layer, and electrolyte decomposition [23,24,28–30]. Nonetheless, perfect reversibility of the capacity was still obtained, and the charge and discharge capacities gradually stabilised in the following 70 cycles.

Cycle stability is another important parameter of the nanosheet-assembled SnO_2-integrated electrode. Figure 6c displays the cycling performance of the nanosheet-assembled SnO_2-integrated electrode with a voltage window of 0.01–1.2 V at a current rate of 1 C. It can be seen that the cycling of the nanosheet-assembled SnO_2-integrated electrode was quite stable, which delivered a high reversible capacity of 637.2 mAh g^{-1} after 70 cycles. It is should be noted that an average Coulombic efficiency of higher than 99% could be obtained up to 70 cycles after the second cycle. In Table S1, we compared the specific capacity of the proposed SnO_2-integrated electrode with the relevant free-standing anode, including SnO_2/CNTs, SnO_2/graphene, and SnO_2 nanoarchitecture on the metal conductive substrate. Given these results, it is expected that synthesising SnO_2-integrated anode assembled by nanosheets could promote the electrochemical performance of SnO_2. Moreover, we also performed the as-synthesised nanoflower, nanosphere, and nanoparticle test as an anode material by the traditional slurry-casting fabrication processes. In comparison, the charge capacity of nanoflower remains 438.5 mAh g^{-1}, outperforming that of the theoretical capacity of graphite (372 mAh g^{-1}). The nanosphere assembled by nanoparticles had an initial capacity of 1383.3 mAh g^{-1}, quickly decreasing to 287.5 mAh g^{-1} after 70 cycles. The nanoparticles without fluorine-doped SnO_2 showed a relatively low capacity of 123.08 mAh g^{-1} after 70 cycles; in particular, the first 10 cyclings deteriorated sharply with the increasing cycle. It is suggested that such nanosheet-assembled SnO_2-integrated electrodes could provide more interconnection between the building blocks and a more stable porous structure due to the effective prevention of dense aggregation of the nanosheets. Such a superior Li-storage performance of nanosheet-assembled SnO_2-integrated electrode should be ascribed to the reasonable structural design on structural engineering to boost synergistic effects. The ultrathin 2D nanoarchitecture blocks provide favourable Li^+ lateral intercalation lengths and more beneficial transport routes for electrolyte ions, and the interconnected 1D micro-fibre core with outstanding metallic conductivity can offer an efficient electron transport pathway along axial orientation to shorten electron transport. More importantly, the metal's remarkable flexibility and high tensile strength endow the hybrid integrated electrode with strong bending and stretchability relative to sintered carbon or graphene hosts. Further, for the electrode realised by the traditional slurry-casting fabrication processes, the better cyclability of nanoflower and nanosphere than that of nanoparticle also supports the key role of fluorine-doped strategy [31,32].

To further evaluate the improved electrochemical performance of the SnO_2-integrated anode, the rate capability at diverse larger current densities was also investigated, which is important for practical applications of LIBs. Benefiting from its unique structure, the nanosheet-assembled SnO_2-integrated electrode revealed an exceptional cycling response to a continuously varying current rate. As displayed in Figure 6d, the representative specific capacities were about 731.17, 677.55, 567.30, and 423.71 mAh g^{-1} at current rates of 1 C, 2 C, 3 C, and 4 C, respectively. The specific capacity slightly decreased as the current density increased, and it could still be maintained a very stable cycling capacity of above 423.71 mAh g^{-1}, corresponding to nearly 100% Coulombic efficiency when the current density was up to 4 C, which was still higher than the theoretical capacity of graphite (372 mAh g^{-1}). More importantly, when the current density was adjusted back to the original current density again, the specific capacity of the SnO_2-integrated electrode still regained the initial reversible value after the high-rate test for the 40 cycles, implying superior stability of the present SnO_2-integrated electrode.

Overall, using microscopic 2D self-assembled materials combined with a macroscopic 2D metal conductor to prepare integrated electrodes is one of the most promising strategies to increase ion and electron transport kinetics toward present LIBs. According to the above results, the presented SnO_2-integrated anode showed significantly improved capacity and rate performance. Clearly, the excellent improvement of electrochemical performance of the nanosheets SnO_2 arrays on the copper micro-fibre electrode can be attributed to two reasons. Firstly, self-supported nanosheets arrays growing directly on a current-collecting substrate represents an attractive nanoarchitecture for LIBs. Such structural feature of thin sheets combined with enriched pores built from the stacking of nanosheets is beneficial to rapid Li^+ intercalation and diffusion of electrolyte into the inner region of the electrode, high electrode–electrolyte contact area, and good stain accommodation. Moreover, each nanosheet has its own contact with the substrate at the bottom, which can ensure every nanosheet participates in the electrochemical reaction and effectively prevent the aggregation of the SnO_2 nanosheet. Secondly, the copper grid was selected as an effective substrate because of its high conductivity, 2D planar structure, and larger mechanical strength, compared with carbon materials. Furthermore, the proposed technique also saves the tedious process of mixing active materials with ancillary materials such as carbon black and polymer binders.

4. Conclusions

In this study, we presented a cost-effective, scalable, and effective approach to fabricate SnO_2 nanosheets cluster arrays directly grown on a 2D-interconnected conductive network with robust mechanical flexibility via a facile hydrothermal route. Such integrated electrodes possess a network configuration, which offers more beneficial transport routes for electrolyte ions and guarantees an intimate contact between active and current collectors. As a result, the SnO_2-integrated electrode with novel nanoarchitecture shows an excellent electrochemical Li-storage performance with a high capacity up to 637.2 mAh g^{-1} at 1 C rate and excellent rate capability of 423.71 mAh g^{-1} at 4 C rate. Such superior cyclic stability and capacity may benefit from the well-designed electrode to boost synergistic effects, which include shortened Li^+ diffusion distance in the 2D nanoarchitecture blocks, sufficient void space among the nanosheets to reduce volume expansion, and a substrate with superior flexibility and robust tensile strength. The presented strategy provides a new synthetic idea for engineering tin-based energy storage systems with high electrochemical performance and robust mechanical flexibility.

Supplementary Materials: The following are available online, Figure S1: XRD pattern of nanoflower SnO_2 assembled by nanosheets without CMF, Figure S2: SEM image of the integrated electrode before ultrasonic treatment, Figure S3: SEM image of SnO_2 samples at 180 °C for 24 h: (a) without NH_4F; (b) F/Sn = 1; (c,d) F/Sn = 2, Table S1: Comparison of electrochemical properties of nanosheet-assembled SnO_2-integrated electrode with other reported Sn-based free-standing anode materials.

Author Contributions: Conceptualisation, X.W. and X.Z.; methodology, X.Z. and Y.W.; investigation, X.W. and X.Z.; writing—original draft preparation, X.Z. and Y.W.; writing—review and editing, X.W. and X.Z. All authors have read and agreed to the published version of the manuscript.

Funding: This research was funded by the Natural Science Foundation of Liaoning Province (Project No: 20180550749), the Foundation of Liaoning Key Laboratory of Chemical Additive Synthesis and Separation (Project No: ZJNK2008), the Doctoral Scientific Research Foundation of Inner Mongolia University for Nationalities (Project No: BS422), and the Opening Foundation from Inner Mongolia Key Lab of Carbon Nanomaterials (Project No: MDK2018050).

Institutional Review Board Statement: Not applicable.

Informed Consent Statement: Not applicable.

Data Availability Statement: Not applicable.

Acknowledgments: X.Z. deeply thanks the West Light Foundation's Visiting Scholar Research Program.

Conflicts of Interest: The authors declare no conflict of interest.

Sample Availability: Samples of the compounds are available from the authors.

References

1. Zhou, G.M.; Li, F.; Cheng, H.M. Progress in flexible lithium batteries and future prospects. *Energy Environ. Sci.* **2014**, *7*, 1307–1338. [CrossRef]
2. Chen, D.; Lou, Z.; Jiang, K.; Shen, G.Z. Device configurations and future prospects of flexible/stretchable lithium-ion batteries. *Adv. Funct. Mater.* **2018**, *28*, 1805596–1805617. [CrossRef]
3. Liu, W.; Song, M.S.; Kong, B.; Cui, Y. Flexible and stretchable energy storage: Recent advances and future perspectives. *Adv. Mater.* **2017**, *29*, 1603436–1603499. [CrossRef]
4. Xia, J.; Zhang, X.; Yang, Y.G.; Wang, X.; Yao, J.N. Electrospinning fabrication of flexible, foldable, and twistable $Sb_2S_3/TiO_2/C$ nanofibre anode for lithium ion batteries. *Chem. Eng. J.* **2021**, *413*, 127400–127408. [CrossRef]
5. Lin, X.P.; Xue, D.Y.; Zhao, L.Z.; Zong, F.Y.; Duan, X.C.; Pan, X.; Zhang, J.M.; Li, Q.H. In-situ growth of 1T/2H-MoS_2 on carbon fibre cloth and the modification of SnS_2 nanoparticles: A three-dimensional heterostructure for high-performance flexible lithium-ion batteries. *Chem. Eng. J.* **2019**, *356*, 483–491. [CrossRef]
6. Zhang, M.; Li, L.H.; Jian, X.L.; Zhang, S.; Shang, Y.Y.; Xu, T.T.; Dai, S.G.; Xu, J.M.; Kong, D.Z.; Wang, Y.; et al. Free-standing and flexible $CNT/(Fe@Si@SiO_2)$ composite anodes with kernel-pulp-skin nanostructure for high-performance lithium-ion batteries. *J. Alloy. Compd.* **2021**, *878*, 160396–160403. [CrossRef]
7. Zhang, P.C.; Cao, M.J.; Feng, Y.; Xu, J.; Yao, J.F. Uniformly growing Co_9S_8 nanoparticles on flexible carbon foam as a free-standing anode for lithium-ion storage devices carbon foam as a free-standing anode for lithium-ion storage devices. *Carbon* **2021**, *182*, 404–412. [CrossRef]
8. Zhou, F.G.; Han, S.C.; Qian, Q.R.; Zhu, Y.F. 3D printing of free-standing and flexible nitrogen doped graphene/polyaniline electrode for electrochemical energy storage. *Chem. Phys. Lett.* **2019**, *728*, 6–13. [CrossRef]
9. Hao, Y.; Wang, C. Free-standing reduced graphene oxide/carbon nanotube paper for flexible sodium-ion battery applications. *Molecules* **2020**, *25*, 1014. [CrossRef]
10. Shao, J.; Yang, Y.; Zhang, X.; Shen, L.; Bao, N. 3D yolk–yhell ytructured Si/void/rGO free-standing electrode for lithium-ion battery. *Materials* **2021**, *14*, 2836. [CrossRef]
11. Li, X.; Bai, Y.; Wang, M.; Wang, G.; Ma, Y.; Huang, Y.; Zheng, J. Dual carbonaceous materials synergetic protection silicon as a high-performance free-standing anode for lithium-ion battery. *Nanomaterials* **2019**, *9*, 650. [CrossRef]
12. Huang, Y.; Li, Y.W.; Huang, R.S.; Ji, J.C.; Yao, J.H.; Xiao, S.H. One-pot hydrothermal synthesis of N-rGO supported Fe_2O_3 nanoparticles as a superior anode material for lithium-ion batteries. *Solid State Ionics* **2021**, *368*, 115693–115700. [CrossRef]
13. Li, B.Q.; Zhao, W.; Yang, Z.; Zhang, C.; Dang, F.; Liu, Y.L.; Jin, F.; Chen, X. A carbon-doped anatase TiO_2-Based flexible silicon anode with high-performance and stability for flexible lithium-ion battery. *J. Power Sources* **2020**, *466*, 228339–228347. [CrossRef]
14. Li, N.; Chen, Z.P.; Ren, W.C.; Li, F.; Cheng, H.M. Flexible graphene-based lithium ion batteries with ultrafast charge and discharge rates. *Proc. Natl. Acad. Sci. USA* **2021**, *109*, 17360–17365. [CrossRef] [PubMed]
15. Liu, S.H.; Wang, Z.Y.; Yu, C.; Wu, H.B.; Wang, G.; Dong, Q.; Qiu, J.S.; Eychmüller, A.; Lou, X.W. A Flexible TiO_2(B)-based battery electrode with superior power rate and ultralong cycle life. *Adv. Mater.* **2013**, *25*, 3462–3467. [CrossRef] [PubMed]
16. Dai, L.; Zhong, X.; Zou, J.; Fu, B.; Su, Y.; Ren, C.; Wang, J.; Zhong, G. Highly ordered SnO_2 nanopillar array as binder-free anodes for long-life and high-rate Li-ion batteries. *Nanomaterials* **2021**, *11*, 1307. [CrossRef] [PubMed]
17. Chen, S.; Wang, M.; Ye, J.; Cai, J.; Ma, Y.; Zhou, H.; Qi, L. Kinetics-controlled growth of aligned mesocrystalline SnO_2 nanorod arrays for lithium-ion batteries with superior rate performance. *Nano Res.* **2013**, *6*, 243–252. [CrossRef]
18. Ding, Y.; Zhou, P.; Han, T.; Liu, J. Environmentally friendly and cost-effective synthesis of carbonaceous particles for preparing hollow SnO_2 nanospheres and their bifunctional Li-storage and gas-sensing properties. *Crystals* **2020**, *10*, 231. [CrossRef]
19. Tran, Q.N.; Kim, I.T.; Park, S.; Choi, H.W.; Park, S.J. SnO_2 Nanoflower–nanocrystalline cellulose composites as anode materials for lithium-ion batteries. *Materials* **2020**, *13*, 3165. [CrossRef]
20. Park, M.S.; Wang, G.X.; Kang, Y.M.; Wexler, D.; Dou, S.X.; Liu, H.K. Preparation and electrochemical properties of SnO_2 nanowires for application in lithium-ion batteries. *Angew. Chem. Int. Ed.* **2007**, *46*, 764–767. [CrossRef]
21. Lou, X.W.; Li, C.M.; Archer, L.A. Designed synthesis of coaxial SnO_2@carbon hollow nanospheres for highly reversible lithium storage. *Adv. Mater.* **2009**, *21*, 2536–2539. [CrossRef]
22. Zhou, X.; Yin, Y.X.; Wan, L.J.; Guo, Y.G. A robust composite of SnO_2 hollow nanospheres enwrapped by graphene as a high-capacity anode material for lithium-ion batteries. *J. Mater. Chem.* **2012**, *22*, 17456–17459. [CrossRef]
23. Li, J.; Zhao, Y.; Wang, N.; Guan, L. A high performance carrier for SnO_2 nanoparticles used in lithium ion battery. *Chem. Commun.* **2011**, *47*, 5238–5240. [CrossRef]
24. Wang, X.Y.; Zhou, X.F.; Yao, K.; Zhang, J.G.; Liu, Z.P. A SnO_2/graphene composite as a high stability electrode for lithium ion batteries. *Carbon* **2011**, *49*, 133–139. [CrossRef]
25. Kim, W.S.; Hwa, Y.; Jeun, J.H.; Sohn, H.J.; Hong, S.H. Synthesis of SnO_2 nano hollow spheres and their size effects in lithium ion battery anode application. *J. Power Sources* **2013**, *225*, 108–112. [CrossRef]
26. Ambalkar, A.A.; Panmand, R.R.; Kawade, U.V.; Sethi, Y.A.; Naik, S.D.; Kulkarni, M.V.; Adhyapak, P.V.; Kale, B.B. Facile synthesis of SnO_2@carbon nanocomposites for lithium-ion batteries. *New J. Chem.* **2020**, *44*, 3366–3374. [CrossRef]
27. Nguyen, T.P.; Kim, I.T. Self-assembled few-layered MoS_2 on SnO_2 anode for enhancing lithium-ion storage. *Nanomaterials* **2020**, *10*, 2558. [CrossRef] [PubMed]

28. Reddy, M.V.; Andreea, L.Y.T.; Ling, A.Y.; Hwee, J.N.C.; Lin, C.A.; Admas, S.; Loh, K.P.; Mathe, M.K.; Ozoemena, K.I.; Chowdari, B.V.R. Effect of preparation temperature and cycling voltage range on molten salt method prepared SnO_2. *Electrochim. Acta* **2013**, *106*, 143–148. [CrossRef]
29. Xu, H.; Chen, J.; Wang, D.; Sun, Z.M.; Zhang, P.G.; Zhang, Y.; Guo, X. Hierarchically porous carbon-coated SnO_2@graphene foams as anodes for lithium ion storage. *Carbon* **2017**, *124*, 565–575. [CrossRef]
30. Han, F.; Li, W.C.; Li, M.R.; Lu, A.H. Fabrication of superior-performance SnO_2@C composites for lithium-ion anodes using tubular mesoporous carbon with thin carbon walls and high pore volume. *J. Mater. Chem.* **2012**, *22*, 9645–9651. [CrossRef]
31. Kwon, C.W.; Campet, G.; Portier, J.; Poquet, A.; Fournes, L.; Labrugere, C.; Jousseaume, B.; Toupance, T.; Choy, J.H.; Subramanian, M.A. A new single molecular precursor route to fluorine-doped nanocrystalline tin oxide anodes for lithium batteries. *J. Inorg. Mater.* **2001**, *3*, 211–214. [CrossRef]
32. Wang, H.K.; Fu, F.; Zhang, F.H.; Wang, H.E.; Kershaw, S.V.; Xu, J.Q.; Sun, S.G.; Rogach, A.L. Hydrothermal synthesis of hierarchical SnO_2 microspheres for gas sensing and lithium-ion batteries applications: Fluoride-mediated formation of solid and hollow structures. *J. Mater. Chem.* **2012**, *22*, 2140–2148. [CrossRef]

Communication

Degradation Investigation of Electrocatalyst in Proton Exchange Membrane Fuel Cell at a High Energy Efficiency

Jie Song [1,*], Qing Ye [1], Kun Wang [2], Zhiyuan Guo [1] and Meiling Dou [2,*]

1   State Key Laboratory of Advanced Transmission Technology, Global Energy Interconnection Research Institute Limited Company, Beijing 102209, China; yeqing@geiri.sgcc.com.cn (Q.Y.); guozhiyuan@geiri.sgcc.com.cn (Z.G.)
2   Beijing Key Laboratory of Electrochemical Process and Technology for Materials, Beijing University of Chemical Technology, Beijing 100029, China; 2018200475@mail.buct.edu.cn
*   Correspondence: songjie@geiri.sgcc.com.cn (J.S.); douml@mail.buct.edu.cn (M.D.)

Abstract: The development of high efficient stacks is critical for the wide spread application of proton exchange membrane fuel cells (PEMFCs) in transportation and stationary power plant. Currently, the favorable operation conditions of PEMFCs are with single cell voltage between 0.65 and 0.7 V, corresponding to energy efficiency lower than 57%. For the long term, PEMFCs need to be operated at higher voltage to increase the energy efficiency and thus promote the fuel economy for transportation and stationary applications. Herein, PEMFC single cell was investigated to demonstrate its capability to working with voltage and energy efficiency higher than 0.8 V and 65%, respectively. It was demonstrated that the PEMFC encountered a significant performance degradation after the 64 h operation. The cell voltage declined by more than 13% at the current density of 1000 mA cm^{-2}, due to the electrode de-activation. The high operation potential of the cathode leads to the corrosion of carbon support and then causes the detachment of Pt nanoparticles, resulting in significant Pt agglomeration. The catalytic surface area of cathode Pt is thus reduced for oxygen reduction and the cell performance decreased. Therefore, electrochemically stable Pt catalyst is highly desirable for efficient PEMFCs operated under cell voltage higher than 0.8 V.

Keywords: proton exchange membrane fuel cell; high energy efficiency; durability; degradation; Pt/C catalyst

Citation: Song, J.; Ye, Q.; Wang, K.; Guo, Z.; Dou, M. Degradation Investigation of Electrocatalyst in Proton Exchange Membrane Fuel Cell at a High Energy Efficiency. *Molecules* **2021**, *26*, 3932. https://doi.org/10.3390/molecules26133932

Academic Editors: Jingqi Guan and Yin Wang

Received: 3 May 2021
Accepted: 4 June 2021
Published: 28 June 2021

Publisher's Note: MDPI stays neutral with regard to jurisdictional claims in published maps and institutional affiliations.

1. Introduction

Proton exchange membrane fuel cells (PEMFCs), as clean and efficient energy conversion devices, have attracted much attention for their promising application in transportation and stationary power plants [1,2]. However, the large-scale commercialization of PEMFCs is still hindered by the high operating costs, mainly due to the low fuel economy and unsatisfied durability [3–6]. Fuel cell energy efficiency is defined as the ratio of output power to the consumed hydrogen enthalpy. For low temperature PEMFCs, the theoretical energy efficiency is about 83%, calculated by dividing the high heating value of hydrogen (286 kJ mol^{-1}) by the Gibbs free energy of fuel cell reaction (237 kJ mol^{-1}). Currently, the favorable operation conditions of PEMFCs are with single cell voltage between 0.65 and 0.7 V, corresponding to energy efficiency lower than 57% [7]. For the long term, PEMFCs need to be operated at higher voltage to increase the energy efficiency and thus promote the fuel economy for transportation and stationary applications. To this end, the U.S. Department of Energy (DOE) has set the technical target of fuel cell systems and stacks for transportation application, with peak energy efficiency increased to 65% in 2020 and 70% in the future [8], respectively. The Japanese government also released the strategic road map for fuel cells by reaching energy efficiency of over 55% in about 2025 and over 65% in the future for stationary applications. Accordingly, PEMFCs should demonstrate the capability of working at a single cell voltage higher than 0.8 V.

The lifetime of PEMFCs is also critical for commercialized applications. For transportation, the lifetime needs to meet the 5000 h target and for stationary power, 40,000 h lifetime is required [9]. In recent years, much efforts have been devoted to the durability investigation of PEMFCs under dynamic operations to mimic the transportation drive cycle [10–15], including idling [16,17], load-cycling [18–20], and start-up/shut-down operation [21–23]. Wang et al. [19] conducted a 900 h duration of accelerated stress test on PEMFC using the drive cycle test protocol developed by Chinese NERC Fuel Cell & Hydrogen Technology, and results showed that the cell performance declined by more than 10% with an average voltage decay rate of 70 μV h^{-1}. It was proposed that the degradation of key materials in PEMFCs is generally caused by the presence of elevated or fluctuating potential at electrodes under dynamic operations [24–28]. Lin et al. [23] adopted an in-situ segmented cell testing technology to analyze the degradation mechanism of the membrane electrode assembly during start-up and shut-down cycles, and results showed that the performance of fuel cells significantly decreased after 1800 cycles.

With the energy efficiency increased, PEMFCs work at higher voltage, and this brings the challenge for their durability. Although there are many investigations on PEMFC durability, the electrochemical stability of the state-of-the-art materials is still unclear when operated with cell voltage higher than 0.8 V. In this work, PEMFC single cell durability was investigated by operating it at an energy efficiency of over 65%. The performance degradation and the material deterioration were investigated. It was shown that the electrode catalyst was corroded electrochemically and the reaction activity was decreased by over 13% after the 64 h operation. Results indicate that the current catalyst support is not durable for highly efficient PEMFCs and more robust Pt support is needed.

2. Results and Discussion

2.1. Durability Test of PEMFC

To mimic the operation of PEMFC at a cell voltage higher than 0.8 V, so that the efficiency exceeds 65%, the operating current density was set at 80 mA cm^{-2}. The durability of PEMFC was then evaluated by constant-current operation for 64 h (Figure 1a). The performance of PEMFC was monitored by measuring the polarization curves every 4 h. In Figure 1b, the cell voltage showed slight fluctuation due to the disturbance caused by the polarization measurement. After each measurement, the voltage decayed from about 0.85 V to 0.81 V, probably due to the re-balance of water content in PEMFC.

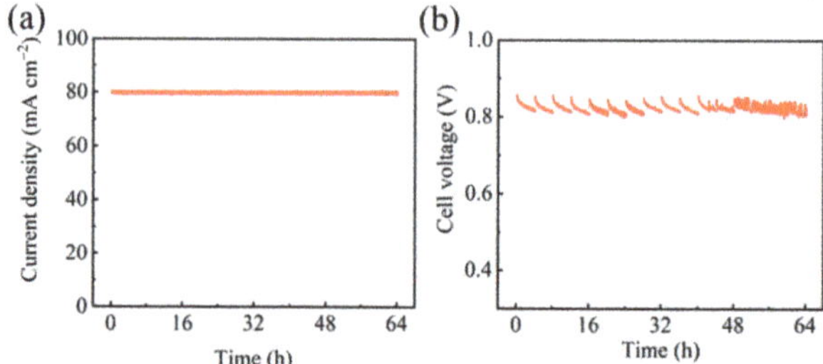

Figure 1. (**a**) Current density versus time and (**b**) cell voltage variation versus time under the current density of 80 mA cm^{-2}.

The polarization curves profiled during the operation were shown in Figure 2 to analyze the whole-range performance. Although the cell voltage at 80 mA cm^{-2} displayed little change after the 64 h operation, PEMFC performance decreased significantly at the current density above 1000 mA cm^{-2}. The cell voltage dropped by 91 mV at 1000 mA cm^{-2}, corresponding to 13.4% performance loss. Considering the short operation duration of only 64 h, the degradation is remarkable, indicating the material deterioration within the cell. Further investigation of the polarization curve indicates that there are increased kinetic, ohmic, and mass transfer loss. The voltage degradation rate was calculated at 300, 1000, and 1400 mA cm^{-2} to demonstrate the performance loss severity at various region. As shown in Figure 3, more significant voltage loss and sharper decay rate were found at higher current density.

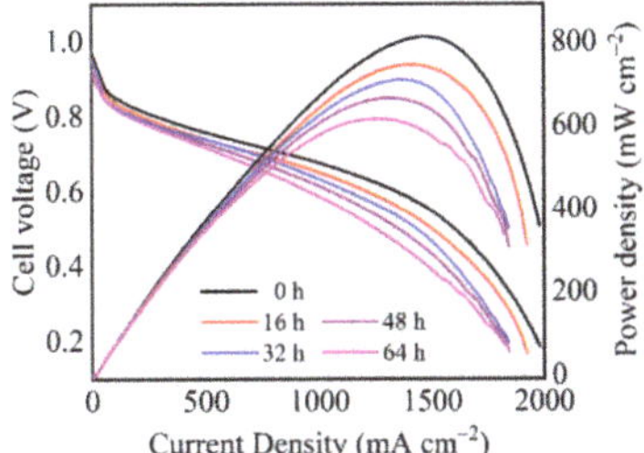

Figure 2. Polarization curves during the durability test.

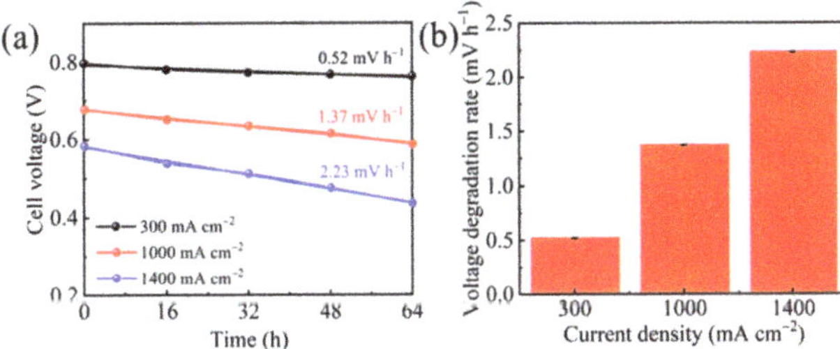

Figure 3. (**a**) Cell voltage versus operation duration and (**b**) the voltage degradation rate at different current densities.

To determine the type of polarization loss, electrochemical impedance spectroscopy (EIS) measurement was carried out at the current density of 150 mA cm^{-2} (Figure 4). By fitting the Nyquist plots, the membrane resistance and charge transfer resistance were obtained. Results show that a slight change was observed for the membrane resistance (Table 1), indicating that the change of proton exchange membrane is negligible during the 64 h test. The charge transfer resistance was found to be increased with the elapse of testing time. After the 64 h test, the charge transfer resistance increased from the initial value of 69.76 mOhm to finally 88.97 mOhm (Table 1), suggesting that the apparent catalytic activity was reduced during the durability test probably due to the loss of Pt active sites.

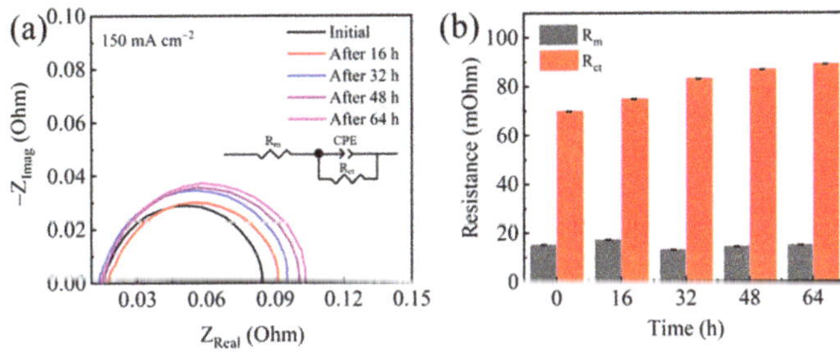

Figure 4. (**a**) Nyquist plots and (**b**) equivalent circuit fitting results at 150 mA cm^{-2} before and after durability test.

Table 1. Fitting results of EIS using equivalent circuit and the ECSA results.

Test Time (h)	Rm (mOhm)	Standard Error	Rct (mOhm)	Standard Error	ECSA ($m^2\ g^{-1}$)	ECSA Loss (%)
0	15.19	0.000102	69.76	0.00046	28.6	0
16	17.18	0.00014	74.80	0.00064	25.2	11.9
32	13.06	0.00011	83.07	0.00067	23.7	17.2
48	14.36	0.00012	86.80	0.00073	19.9	30.6
64	14.93	0.000096	88.97	0.00056	18.2	36.6

R_m: membrane resistance; R_{ct}: charge transfer resistance; CPE: constant phase angle element.

To investigate the degradation of Pt/C electrocatalyst, we performed the CV measurement with the scan rate of 50 mV s^{-1} for PEMFC cathode to estimate the cathode electrochemical surface areas (ECSAs, unit: $m^2\ g^{-1}_{Pt}$). The value of ECSA indicates the amount of available active sites in the catalyst layer for the oxygen reduction reaction in the cathode. Large ECSA is favorable for a high performance cathode. The ECSAs were calculated by measuring the hydrogen desorption charge integrated from the corresponding potential region in CV curves assuming 0.21 mC cm^{-2} as the charge that is required for oxidizing the hydrogen monolayer on Pt. For comparison, all the calculated ECSAs at different test times are normalized by the initial value. As shown in Figure 5, the corresponding current densities for both the desorption of adsorbed hydrogen and the reduction of Pt-Ox both gradually decrease with the elapse of the testing time, indicating the reduction of ECSAs during the durability test. Therefore, the effective Pt active sites in the cathode were decreased under the high-voltage operation condition. A further investigation indicates that the normalized ECSAs can be linearly fitted with the operation duration. As shown in Figure 6, an average degradation rate of 0.57% h^{-1} was deduced for cathode ECSA.

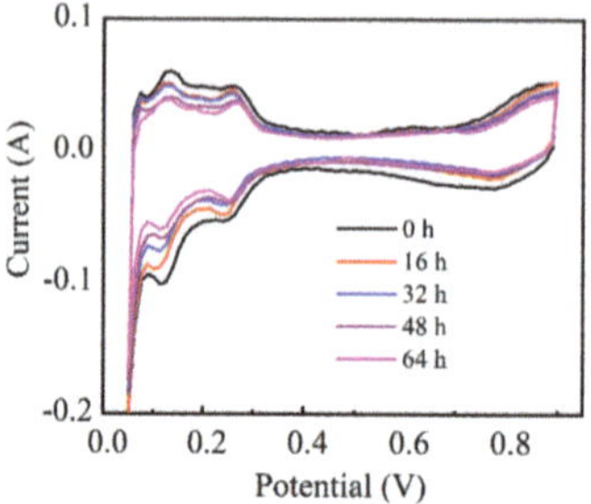

Figure 5. CV curves profiled during the 64 h durability testing.

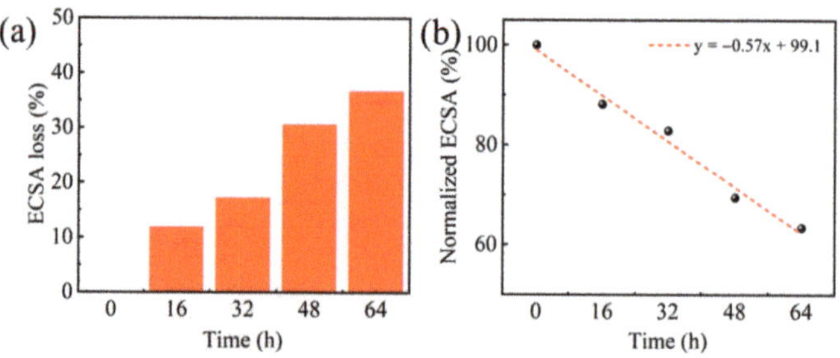

Figure 6. (a) Normalized ECSA loss measured during the 64 h test; (b) Linear fitting of normalized ECSA loss with operation duration.

2.2. Physicochemical Characterization of the Cathodei Catalyst Layer

The durability of the catalyst layer is critical for the lifetime of the fuel cell because the catalyst layer is heart component in membrane electrode assembly for the fuel cell stack. Significant degradation of the catalyst layer leads to the increase in the kinetic polarization loss, irreversibly affecting the durability of fuel cells [29]. Therefore, physicochemical characterizations were carried out to investigate the degradation mechanism of the catalyst layer under the durability test. SEM images show that the cathodic catalyst layer displays a relatively smooth surface distributed with the Pt/C catalyst before the test (Figure 7). After 64 h operation, the surface of the cathodic catalyst layer becomes rougher than the initial catalyst layer, with some collapsed pores in some regions. It implies that the operation at a high voltage has a negative effect on the microstructure of cathodic catalyst layer due to the presence of high potential, and also indicates that the current key materials in fuel cells are not durable.

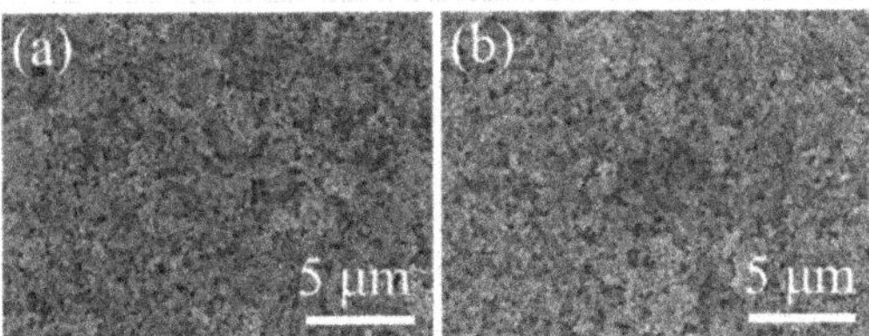

Figure 7. SEM images of the cathodic catalyst layer surface (**a**) before and (**b**) after 64 h durability test.

TEM characterization was performed to observe the microstructure change of the cathodic catalyst layer before and after the test (Figure 8). TEM images show that the Pt nanoparticles in the Pt/C electrocatalyst mostly exhibit a spherical morphology with an average particle size of approximately 4.0 nm before the test. After the 64 h test, Pt nanoparticles aggregate seriously with an enlarged particle size of approximately 5.4 nm. The obvious growth of Pt nanoparticles is consistent with the change of ECSAs and form CV curves, further confirming that the Pt utilization decreases under such high working cell voltage. Furthermore, the morphology of these Pt nanoparticles changes to be irregular and complex in comparison to their initial morphology, showing significant agglomeration of Pt nanoparticles after the 64 h test. This results reveals that a serious degradation of the Pt/C catalyst in the cathode occurs, which is due to the lasting presence of high potential above 0.8 V. It probably causes corrosion of the carbon support in the Pt/C catalyst, and then leads to the loss of Pt reactive sites, that is, Pt nanoparticles fall off from the carbon surface and agglomerate with an enlarged particle size. Therefore, the effective Pt reactive sites are reduced and ultimately result in a decline in the electrocatalytic activity and thus a decrease in the fuel cell performance.

To investigate the change of surface element composition and content on the cathodic catalyst layer, XPS measurement was conducted for the sample before and after the durability test (Figure 9a). XPS spectra show the presence of C, Pt, F, S, and O on the surface of the cathodic catalyst layer for both before and after the 64 h test, indicating that the main component for the cathodic catalyst layer is the Pt/C catalyst and Nafion ionomer. Further investigations showed that the Pt content decreases significantly after the 64 h test, from 4.75 at% to 3.72 at% (Table 2). The decrease of Pt content after the durability test further indicates that Pt nanoparticles might detach from the carbon support due to the corrosion of carbon under high potential, which is consistent with the TEM results.

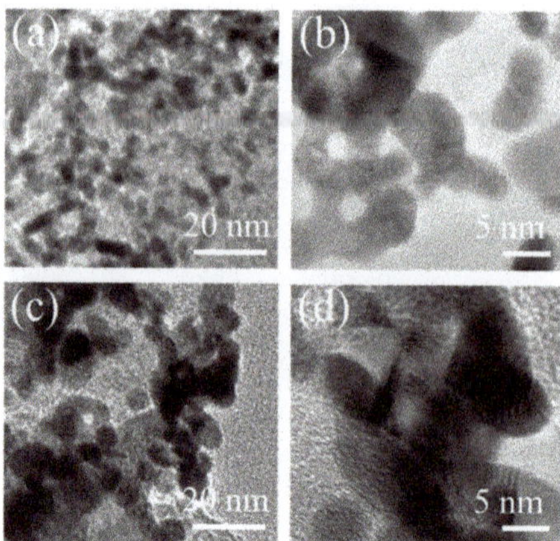

Figure 8. TEM images of cathode Pt/C catalyst before (**a,b**) and after (**c,d**) the 64 h test at a current density of 80 mA cm^{-2}.

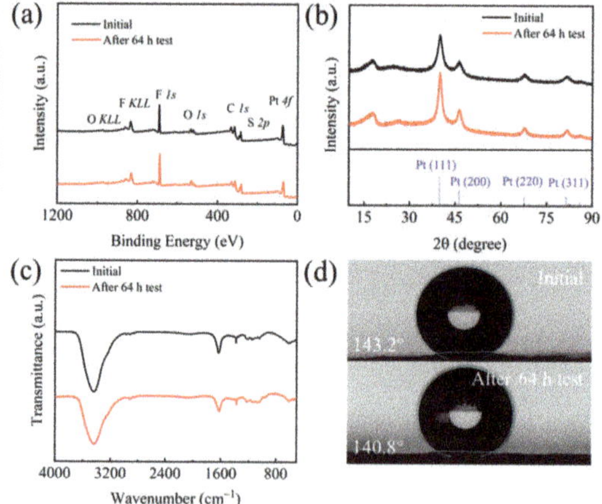

Figure 9. (**a**) XPS survey spectra, (**b**) XRD patterns, and (**c**) FTIR spectra of the cathodic catalyst layer before and after the 64 h test. (**d**) Contact angle test of the cathodic gas diffusion layer before and after the 64 h test.

Table 2. Elemental composition and content before and after the 64 h test.

Element	C at%	O at%	F at%	Pt at%	S at%
Initial	49.05	8.24	36.21	4.75	1.75
After 64 h test	49.58	9.48	35.37	3.72	1.84

XRD patterns were conducted for the membrane electrode assembly before and after the 64 h test (Figure 8b). Results show the diffraction peaks at 39.76°, 46.24°, 67.45°, and 81.29° are assigned to the (111), (200), (220) and (311) planes, respectively, of face-centered cubic (fcc) structured Pt (JCPDS No. 04-0802). The peaks at 18.10° and 26.60° are corresponding to the polytetrafluoroethylene (PTFE) (100) and C (002) plane, respectively, which also confirms the main component of Nafion ionomer and Pt/C catalyst in the catalyst layer. After the 64 h test, no other crystal structure was observed in the membrane electrode assembly. However, the Pt (111) peak becomes sharper than the initial sample, indicating the larger crystalline size of Pt after the durability test, which is in agreement with the TEM results. FTIR test shows that the peaks at 1383, 1231, and 1158 are assigned to the stretching vibrations of S=O bond, asymmetric C-F bond, and symmetric C-F bond (Figure 8c), respectively, confirming the presence of Nafion ionomer. After the 64 h test, no obvious change was observed for the infrared characteristic peaks, indicating that the Nafion ionomer in the catalytic layer did not change obviously. The contact angle test results indicated that the contact angle of the cathodic gas diffusion layer changes negligibly after the 64 h test (from 143.2° to 140.8°) (Figure 8d), suggesting that the durability test did not affect the hydrophobicity of the gas diffusion layer.

3. Materials and Methods

3.1. Fuel Cell Test

A PEMFC with an active area of 5 cm^2 was tested by a Scribner 850e fuel cell test station from Hephas Energy Co., Ltd. The catalyst-coated membrane (CCM) sample purchased from the Shaoxing Junji Energy Technology Co., Ltd., Shaoxing, China, in 2019 was fabricated by coating the commercial Pt/C catalyst on Nafion 212 membrane and then combined with two gas diffusion layers to form the membrane electrode assembly. The Pt loading in the anode and cathode catalyst layer is 0.2 and 0.6 mg cm^{-2}, respectively, and the fuel cell temperature was set at 80 °C. The flow of hydrogen (H_2) and the air was controlled at 200 and 800 mL min^{-1}, respectively, with the back pressure of 100 kPa and gas relative humidity (RH) of 100% for both anode and cathode. Electrochemical impedance spectroscopy (EIS) test was performed by applying a sine wave distortion (AC perturbation) of 10% DC current amplitude under the galvanostatic mode (frequency range: 10 kHz–100 mHz) to record the impedance spectra with air in cathode and H_2 in the anode. Cyclic voltammetry (CV) test was conducted to characterize the change in the loss of Pt sites with N_2 in cathode and H_2 in anode at the scan rate of 50 mV s^{-1}.

3.2. Material Characterization

The morphology of the sample was characterized by the scanning electron microscope (SEM, JEOL FE-JSM-6701F) and the transmission electron microscopy (TEM, JEM-2100, JEM-2010F, JEOL, Tokyo, Japan). The surface element composition and content was characterized by X-ray photoelectron spectroscopy (XPS) (Thermo Fisher Scientific ESCALAB 250) using an Al Kα source. The crystal structure of the sample was characterized using a D/max-2500 X-ray diffraction diffractometer (XRD) equipped with Cu Kα radiation. Fourier transform infrared (FTIR) spectra were recorded on a Nicolet 6700 FTIR spectrophotometer in the range 400–4000 cm^{-1}.

4. Conclusions

In summary, we have evaluated the durability behavior of a single PEMFC at a high energy efficiency (over 65%) that operates at a voltage higher than 0.8 V and elucidated the material degradation as characteristic. The cell performance decays with a rate of ~1.37 mV h^{-1} at a current density of 1000 mA cm^{-2} during the 64 h test. The decisive factor for the durability deterioration of PEMFC is the degradation of Pt/C catalyst in cathode, showing an obvious growth of Pt nanoparticles with significant Pt aggregation under this operation due to the lasting presence of high potential. This work indicates that the current catalyst is not sufficiently durable for highly efficient PEMFCs, and thus, the exploration of a robust catalyst that is more resistant to a high potential is significantly urgent.

Author Contributions: Conceptualization, J.S. and Q.Y.; methodology, M.D.; investigation, Z.G.; data curation, K.W.; writing-original draft preparation, K.W. and M.D.; writing-review and editing, M.D. All authors have read and agreed to the published version of the manuscript.

Funding: This work was supported by the State Grid Corporation Technology Support Project: Research on Key Technology of Membrane Electrode of Proton Exchange Membrane Fuel Cell Suitable for Variable Load (Contract No. 2018-2020-SGGR0000DLJS1800814).

Institutional Review Board Statement: Not applicable.

Informed Consent Statement: Not applicable.

Conflicts of Interest: The authors declare no conflict of interest.

References

1. Gamburzev, S.; Appleby, A. Recent progress in performance improvement of the proton exchange membrane fuel cell (PEMFC). *J. Power Sources* **2002**, *107*, 5–12. [CrossRef]
2. Zhang, X.; Zhang, T.; Chen, H.; Cao, Y. A review of online electrochemical diagnostic methods of on-board proton exchange membrane fuel cells. *Appl. Energy* **2021**, *286*, 116481. [CrossRef]
3. Wu, J.; Yuan, X.Z.; Martin, J.J.; Wang, H.; Zhang, J.; Shen, J.; Wu, S.; Merida, W. A review of PEM fuel cell durability: Degradation mechanisms and mitigation strategies. *J. Power Sources* **2008**, *184*, 104–119. [CrossRef]
4. Pan, M.; Pan, C.; Li, C.; Zhao, J. A review of membranes in proton exchange membrane fuel cells: Transport phenomena, performance and durability. *Renew. Sustain. Energy Rev.* **2021**, *141*, 110771. [CrossRef]
5. Shao, Y.; Yin, G.; Gao, Y. Understanding and approaches for the durability issues of Pt-based catalysts for PEM fuel cell. *J. Power Sources* **2007**, *171*, 558–566. [CrossRef]
6. De Bruijn, F.D.; Dam, V.A.T.; Janssen, G.J.M. Review: Durability and Degradation Issues of PEM Fuel Cell Components. *Fuel Cells* **2008**, *8*, 3–22. [CrossRef]
7. Zhou, X.; Yang, Y.; Li, B.; Zhang, C. Advanced Reversal Tolerant Anode in Proton Exchange Membrane Fuel Cells: Study on the Attenuation Mechanism during Fuel Starvation. *ACS Appl. Mater. Interfaces* **2021**, *13*, 2455–2461. [CrossRef]
8. DOE Technical Targets for Fuel Cell Systems and Stacks for Transportation Applications. Available online: https://www.energy.gov/eere/fuelcells/doe-technical-targets-fuel-cell-systems-and-stacks-transportation-applications (accessed on 3 May 2021).
9. Ren, P.; Pei, P.; Li, Y.; Wu, Z.; Chen, D.; Huang, S. Degradation mechanisms of proton exchange membrane fuel cell under typical automotive operating conditions. *Prog. Energy Combust. Sci.* **2020**, *80*, 100859. [CrossRef]
10. Jahnke, T.; Futter, G.; Latz, A.; Malkow, T.; Papakonstantinou, G.; Tsotridis, G.; Schott, P.; Gérard, M.; Quinaud, M.; Quiroga, M.; et al. Performance and degradation of Proton Exchange Membrane Fuel Cells: State of the art in modeling from atomistic to system scale. *J. Power Sources* **2016**, *304*, 207–233. [CrossRef]
11. Curtin, D.E.; Lousenberg, R.D.; Henry, T.J.; Tangeman, P.C.; Tisack, M.E. Advanced materials for improved PEMFC performance and life. *J. Power Sources* **2004**, *131*, 41–48. [CrossRef]
12. Pei, P.; Chang, Q.; Tang, T. A quick evaluating method for automotive fuel cell lifetime. *Int. J. Hydrog. Energy* **2008**, *33*, 3829–3836. [CrossRef]
13. Kang, J.; Kim, J. Membrane electrode assembly degradation by dry/wet gas on a PEM fuel cell. *Int. J. Hydrog. Energy* **2010**, *35*, 13125–13130. [CrossRef]
14. Chu, T.; Zhang, R.; Wang, Y.; Ou, M.; Xie, M.; Shao, H.; Yang, D.; Li, B.; Ming, P.; Zhang, C. Performance degradation and process engineering of the 10 kW proton exchange membrane fuel cell stack. *Energy* **2021**, *219*, 119623. [CrossRef]
15. Pei, P.C.; Chen, H.C. Main factors affecting the lifetime of Proton Exchange Membrane fuel cells in vehicle applications: A review. *Appl. Energy* **2014**, *125*, 60–75. [CrossRef]
16. Wu, J.; Yuan, X.-Z.; Martin, J.J.; Wang, H.; Yang, D.; Qiao, J.; Ma, J. Proton exchange membrane fuel cell degradation under close to open-circuit conditions: Part I: In situ diagnosis. *J. Power Sources* **2010**, *195*, 1171–1176. [CrossRef]
17. Yuan, X.-Z.; Zhang, S.; Wang, H.; Wu, J.; Sun, J.C.; Hiesgen, R.; Friedrich, K.A.; Schulze, M.; Haug, A. Degradation of a polymer exchange membrane fuel cell stack with Nafion® membranes of different thicknesses: Part I, in situ diagnosis. *J. Power Sources* **2010**, *195*, 7594–7599. [CrossRef]
18. Wang, G.; Huang, F.; Yu, Y.; Wen, S.; Tu, Z. Degradation behavior of a proton exchange membrane fuel cell stack under dynamic cycles between idling and rated condition. *Int. J. Hydrog. Energy* **2018**, *43*, 4471–4481. [CrossRef]
19. Wang, C.; Zhao, Q.; Zhou, X.; Wang, J.; Tang, Y. Degradation characteristics of membrane electrode assembly under drive cycle test protocol. *Int. J. Green Energy* **2019**, *16*, 789–795. [CrossRef]
20. Garcia-Sanchez, D.; Morawietz, T.; da Rocha, P.G.; Hiesgen, R.; Gazdzicki, P.; Friedrich, K. Local impact of load cycling on degradation in polymer electrolyte fuel cells. *Appl. Energy* **2020**, *259*, 114210. [CrossRef]
21. Zhang, T.; Wang, P.; Chen, H.; Pei, P. A review of automotive proton exchange membrane fuel cell degradation under start-stop operating condition. *Appl. Energy* **2018**, *223*, 249–262. [CrossRef]
22. Mittermeier, T.; Weiß, A.; Hasché, F.; Hübner, G.; Gasteiger, H.A. PEM Fuel Cell Start-up/Shut-down Losses vs Temperature for Non-Graphitized and Graphitized Cathode Carbon Supports. *J. Electrochem. Soc.* **2016**, *164*, F127–F137. [CrossRef]
23. Lin, R.; Xiong, F.; Tang, W.; Técher, L.; Zhang, J.; Ma, J. Investigation of dynamic driving cycle effect on the degradation of proton exchange membrane fuel cell by segmented cell technology. *J. Power Sources* **2014**, *260*, 150–158. [CrossRef]
24. Owejan, J.E.; Yu, P.T.; Makharia, R. Mitigation of Carbon Corrosion in Microporous Layers in PEM Fuel Cells. *ECS Trans.* **2007**, *11*, 1049–1057. [CrossRef]
25. Ferreira, P.J.; la O', G.J.; Shao-Horn, Y.; Morgan, D.; Makharia, R.; Kocha, S.; Gasteiger, H.A. Instability of Pt/C Electrocat-alysts in Proton Exchange Membrane Fuel Cells. *J. Electrochem. Soc.* **2005**, *152*, A2256. [CrossRef]

26. Shao, J.; Huang, H.; Lu, L.; Pei, P. An experimental study on the performance of automotive PEMFC under typical conditions. *Automot. Eng.* **2007**, *29*, 566–569.
27. Vengatesan, S.; Panha, K.; Fowler, M.W.; Yuan, X.-Z.; Wang, H. Membrane electrode assembly degradation under idle conditions via unsymmetrical reactant relative humidity cycling. *J. Power Sources* **2012**, *207*, 101–110. [CrossRef]
28. Darling, R.M.; Meyers, J.P. Mathematical Model of Platinum Movement in PEM Fuel Cells. *J. Electrochem. Soc.* **2005**, *152*, A242–A247. [CrossRef]
29. Zhang, S.; Yuan, X.-Z.; Hin, J.N.C.; Wang, H.; Wu, J.; Friedrich, K.A.; Schulze, M. Effects of open-circuit operation on membrane and catalyst layer degradation in proton exchange membrane fuel cells. *J. Power Sources* **2010**, *195*, 1142–1148. [CrossRef]

Article

Fabrication of Highly Textured 2D SnSe Layers with Tunable Electronic Properties for Hydrogen Evolution

Qianyu Zhou [1,2,3,†], Mengya Wang [1,2,3,†], Yong Li [1,2,3], Yanfang Liu [1,2,4], Yuanfu Chen [2,4,*], Qi Wu [1,2,3,*] and Shifeng Wang [1,2,3,*]

1 Department of Physics, and Innovation center of Materials for Energy and Environment Technologies, College of Science, Tibet University, Lhasa 850000, China; zhouqianyu@utibet.edu.cn (Q.Z.); wmy1430140113@sina.com (M.W.); xzuliyong@utibet.edu.cn (Y.L.); liuyanfang@utibet.edu.cn (Y.L.)
2 Institute of Oxygen Supply, Center of Tibetan Studies (Everest Research Institute), Tibet University, Lhasa 850000, China
3 Key Laboratory of Cosmic Rays (Tibet University), Ministry of Education, Lhasa 850000, China
4 School of Electronic Science and Engineering, and State Key Laboratory of Electronic Thin Films and Integrated Devices, University of Electronic Science and Technology of China, Chengdu 610054, China
* Correspondence: yfchen@uestc.edu.cn (Y.C.); wuqi@utibet.edu.cn (Q.W.); wsf365@163.com (S.W.)
† The authors contributed equally to this work.

Abstract: Hydrogen is regarded to be one of the most promising renewable and clean energy sources. Finding a highly efficient and cost-effective catalyst to generate hydrogen via water splitting has become a research hotspot. Two-dimensional materials with exotic structural and electronic properties have been considered as economical alternatives. In this work, 2D SnSe films with high quality of crystallinity were grown on a mica substrate via molecular beam epitaxy. The electronic property of the prepared SnSe thin films can be easily and accurately tuned in situ by three orders of magnitude through the controllable compensation of Sn atoms. The prepared film normally exhibited p-type conduction due to the deficiency of Sn in the film during its growth. First-principle calculations explained that Sn vacancies can introduce additional reactive sites for the hydrogen evolution reaction (HER) and enhance the HER performance by accelerating electron migration and promoting continuous hydrogen generation, which was mirrored by the reduced Gibbs free energy by a factor of 2.3 as compared with the pure SnSe film. The results pave the way for synthesized 2D SnSe thin films in the applications of hydrogen production.

Keywords: SnSe; 2D materials; hydrogen evolution; water splitting; DFT calculations; defect engineering

Citation: Zhou, Q.; Wang, M.; Li, Y.; Liu, Y.; Chen, Y.; Wu, Q.; Wang, S. Fabrication of Highly Textured 2D SnSe Layers with Tunable Electronic Properties for Hydrogen Evolution. *Molecules* **2021**, *26*, 3319. https://doi.org/10.3390/molecules26113319

Academic Editors: Jingqi Guan and Yin Wang

Received: 11 May 2021
Accepted: 28 May 2021
Published: 1 June 2021

Publisher's Note: MDPI stays neutral with regard to jurisdictional claims in published maps and institutional affiliations.

1. Introduction

With the increase in CO_2 emissions due to the use of traditional energy sources such as coal-fired electricity and oil-powered cars, the global warming leading to the sea level rise, glaciers melting, and frequent extreme weather has become an issue of increasing concern all over the world. To address this challenge, the concept of "carbon neutrality" was coined in 2005 [1,2], which referred to an equilibrium between the CO_2 emissions in the atmosphere and the removal or capture of CO_2 from the atmosphere generating net zero emissions. Several effective strategies have been proposed to reduce the amount of CO_2 emissions, such as planting more trees, encouraging use of renewable energy sources, improving energy efficiency, and developing clean transportation. Hydrogen energy is such a kind of clean energy source with high efficiency and environmental benefits, whose "fuel" is hydrogen and/or hydrogen-containing compounds. Water splitting is considered to be the most promising pathway for hydrogen production, as it is recyclable, clean, and abundant. According to the thermodynamics of water splitting, the energy required to break one mole of water is 237 kJ [3,4], and the potential needed is at least 1.23 V [5,6]. To dissociate water molecules into hydrogen and oxygen by means of photoelectrochemical catalysis, various semiconductor materials have been used as catalysts, such as TiO_2 [7,8], $SrTiO_3$ [9,10], $BiVO_4$ [10–12], and chalcogenide compounds [13,14].

To match the hydrogen reduction potential and water oxidation potential, wide-bandgap semiconductors are usually used as catalysts, which absorb only the sunlight in the UV range that comprises only a small amount of the solar irradiation energy. Other factors limiting the splitting efficiency include low charge separation efficiency, fast electron–hole recombination, and slow kinetics of the water redox reaction [4]. Although heterogenous junctions and Z-scheme structures have been designed to boost water splitting efficiency [10,15–18], most of the conventional catalysts still exhibit insufficiency because of the long migration path of the photogenerated carriers and lack of reactive sites.

Two-dimensional (2D) materials have attracted great interest in catalytic applications due to their anisotropic physical and electronic properties, high carrier mobility, tunable energy bandgaps, and high surface-to-bulk ratio facilitating enrichment of reactive sites and shortening the migration distance of carriers. Very recently, various 2D materials, such as graphene [19], graphitic carbon nitride (g-C_3N_4) [20], MoS_2 [21], $MoSe_2$ [22], WSe_2 [23], and oxosulfide [24], have been synthesized for the oxygen evolution reaction (OER) and hydrogen evolution reaction (HER) activities. Belonging to the family of 2D transition metal chalcogenides, tin monoselenide (SnSe) has also received researchers' attention due to its simplicity in structure and fabrication, inexpensiveness of constituent sources, superior performance for catalytic activity, and compatibility with diverse thin film preparation techniques. It has been studied as a catalyst for CO_2 reduction [25]. However, there are few studies on the hydrogen generation using SnSe as the catalyst. SnSe crystalizes in an orthorhombic unit cell, in which atoms are strongly connected by covalent bonds within the layer whereas weak van der Waals interactions occur between the layers [26,27]. This unique structure enables SnSe to easily achieve 2D/2D or 2D/3D stacking forming heterostructures for the catalytic activity.

In this paper, layered SnSe films with high quality of crystallinity were successfully grown on mica substrates. A full width at half maximum (FWHM) of the XRD rocking curve was achieved as narrow as $0.121°$ on the SnSe (004) plane, which might be the best value ever reported and suggests highly textured growth orientation along its c-axis with excellent crystallization quality. Electrical measurements revealed that the films showed a p-type conductivity due to Sn vacancies in the film with a carrier mobility as high as $34\ cm^2/(Vs)$ and a sheet resistivity of $1.5 \times 10^4\ \Omega/square$. The vacant defects can be effectively tuned by a separate elemental tin compensation source, tailoring resistivity in the range of three orders of magnitude. First-principle calculations using the Vienna Ab Initio Simulation Package (VASP) revealed that the presence of Sn vacancies in the SnSe film reduced the Gibbs free energy by a factor of 2.3 as compared with the pure SnSe. It was explained that Sn vacancies on the surface of layers provide more reactive sites and favor separation and transportation of photogenerated electrons, facilitating the continuous hydrogen evolution reaction. Our results pave the way to explore such novel 2D materials as economical alternatives to the expensive platinum-based catalysts for hydrogen generation.

2. Materials and Methods

2.1. Preparation and Characterization of Materials

SnSe films were deposited on mica substrates using the molecular beam epitaxy (MBE) technique, in which compound SnSe pieces with 5N purity (purchased from American Elements, Los Angeles, CA, USA) were used as the evaporation source and tin pellets with 5N purity (purchased from American Elements, Los Angeles, CA, USA) were employed as the compensation source, loaded in separate K-cells. The substrate temperature was kept at about $250\ °C$ for film growth, while the SnSe source was heated up to $450\ °C$ for evaporation. The temperature of the tin source varied in the range between $700\ °C$ and $800\ °C$ to compensate the Sn vacancies and regulate the electrical property of the SnSe film.

The crystal phase of the prepared SnSe films was examined by XRD, which was performed using a Rigaku Smartlab 9 kW X-ray diffractometer with the incident wavelength of 1.5406 Å (Rigaku Corporation, Tokyo, Japan). Atomic force microscopy (AFM) was employed to depict the film surface morphology using a Bruker NanoScope 8 (Billerica, MA, USA) in the tapping mode. Ultraviolet photoelectron spectroscopy (UPS) was used to determine the work function of the prepared SnSe film, which was recorded on an EscaLab 250 spectrometer (Thermo Fisher Scientific, Waltham, MA, USA) with an energy step of 20 meV, using He(I) radiation (hν = 21.22 eV) as the UV source. A Bio-Rad 5500 Hall system (Hercules, CA, USA) equipped with a permanent magnet with a magnetic flux density of 0.32 T was used to determine the electrical property applying the four-probe van der Pauw method.

2.2. Computation Details

Density functional theory (DFT) computations were performed using the plane-wave basis set in the VASP with the projector augment wave (PAW) method [28,29]. Exchange and correlation effects for the structural relaxation were approximated by generalized gradient approximation (GGA) utilizing the Perdew–Burke–Ernzerhof (PBE) functional [30,31]. The Grimme custom method for DFT-D3 correction was employed to precisely depict the impacts of van der Waals interactions [32,33]. The HSE06 (Heyd–Scuseria–Ernzerhof) functional was utilized for electronic structure computations because the PBE functional typically underestimates the bandgap value [28]. The cutoff energy was set to be 500 eV for the plane-wave basis set. The Brillouin zone (BZ) was sampled using a $8 \times 7 \times 3$ k-point Monkhorst–Pack sampling grid for the bulk SnSe and a $2 \times 2 \times 1$ grid for the SnSe monolayer. The convergence criteria of energy and force were 1×10^{-5} eV and 0.01 eV/Å, respectively. A vacuum layer of 15 Å was added along the c-axis of the SnSe monolayer to avoid the impact of the periodic layer.

To obtain further insights into the HER performance of SnSe, DFT simulations were conducted to compute the free energy (ΔG_{H*}) of H adsorption, which is usually employed as a key indicator for HER activity. To find which surface is more conducive to the HER, an SnSe monolayer surface and a surface with one Sn defect introduced were constructed as shown in Figure 1.

The adsorption of H atoms on the surface was studied, and the hydrogen chemisorption energy was computed as follows:

$$\Delta E_{H*} = E_{slab+H} - E_{slab,clean} - \frac{1}{2}E_{H_2,gas} \tag{1}$$

where E_{slab+H} stands for the total energy of the adsorbed hydrogen atom on the surface, $E_{slab,clean}$ is the calculated energy of a clean surface, and $E_{H_2,gas}$ is the total energy of an H_2 molecule in the gaseous state.

The free energy of the systems can be expressed as follows:

$$\Delta G_{H*} = \Delta E_{H*} + \Delta E_{ZPE} - T\Delta S_H \tag{2}$$

where ΔG_{H*}, ΔE_{H*}, ΔE_{ZPE}, and ΔS_H denote the free energy of the system, the aforementioned adsorption energy, the zero-point energy change, and the entropy change between adsorbed hydrogen and hydrogen in the gaseous state at standard conditions, respectively; ΔS_H is roughly equal to $\frac{1}{2}\Delta S_{H_2}$, where ΔS_{H_2} is the entropy of an isolated H_2 molecule in the gaseous state at standard conditions; therefore, the value of $T\Delta S_H$ is approximately -0.2 eV. ΔE_{ZPE} can be described as follows:

$$\Delta E_{ZPE} = E_{ZPE}^H - \frac{1}{2}E_{ZPE}^{H_2} \tag{3}$$

where E_{ZPE}^H and $E_{ZPE}^{H_2}$ represent the zero-point energy of an adsorbed hydrogen atom as well as of a hydrogen molecule in the gaseous state, respectively.

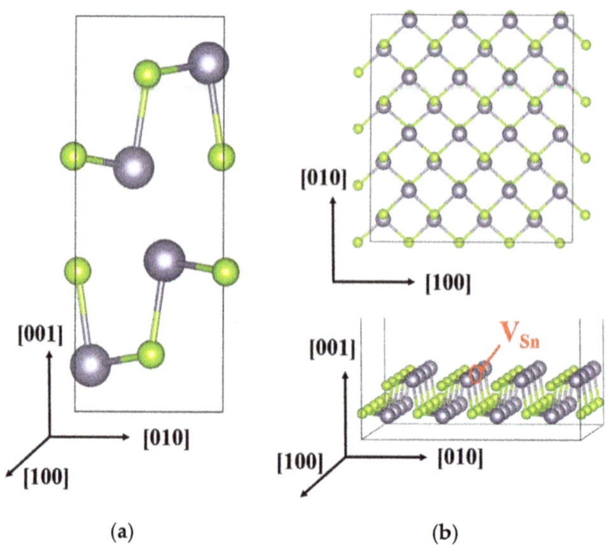

Figure 1. The structure of SnSe. (**a**) Unit cell of bulk SnSe. (**b**) Top view and perspective view of the constructed SnSe supercell. Navy-colored balls represent Sn atoms and green balls represent Se atoms. The red circle indicates an Sn vacancy.

3. Results and Discussion

Being a 2D layered material, high-quality SnSe layers can be obtained on mica substrates, which also belong to the class of 2D materials providing a chemically inert, atomically flat, and electrically insulating surface [26]. As shown in Figure 2, only the diffraction peaks at 15.3°, 30.9°, 47.1°, and 64.4° originating from the SnSe (002) family planes emerge in the XRD pattern, suggesting a highly textured growth along its c-axis, namely perfect layer-by-layer stacking. In addition, an in-plane phi scan of the SnSe (016) plane with respect to the SnSe (001) plane was conducted by tilting the sample at an angle of $\chi = 24.71°$ and setting the incident x-ray angle of $2\theta = 52.17°$ as shown in Figure S1 in the Supplementary Materials. As a result, the lattice constants of SnSe can be derived to be a = 4.42 Å, b = 4.19 Å, c = 11.57 Å. The XRD pattern and the calculated crystal parameters are in excellent agreement with the JCPDS database (No. 1089–0236). The inset in Figure 2 describes the XRD rocking curve carried out on the SnSe (004) plane with respect to a mica substrate. The narrow peak demonstrates an FWHM as small as 0.121°, which might be the narrowest value ever reported and indicates excellent quality of SnSe crystallinity [26,27].

The AFM morphology is displayed in Figure 3. The root-mean-square (RMS) roughness was calculated to be 1.03 nm (on a two-micron scale), indicating a very flat SnSe (001) surface. Orthorhombic terrace-like features emerged with the size of ~500 nm and a height of 0.68 nm on the average, signifying the monolayer thickness, which is consistent with the value obtained from the XRD data in Figure 2.

The Hall measurements revealed that the prepared SnSe film exhibited a p-type conductivity due to Sn deficiency during the crystallization, introducing acceptor states in the film. The Hall mobility was measured to be 34 $cm^2/(V\cdot S)$ at the sheet resistivity of 1.5×10^4 Ω/square. Fortunately, Sn defects can be effectively compensated by adding elemental tin atoms simultaneously during film growth and the amount of compensating tin atoms incorporated in the film can be precisely regulated by varying the elemental tin source temperature. As shown in Figure S2 in the Supplementary Materials, sheet resistivity can be adjusted three orders of magnitude larger than that without Sn compensation, making the SnSe crystal nearly perfect with the least Sn vacant defects.

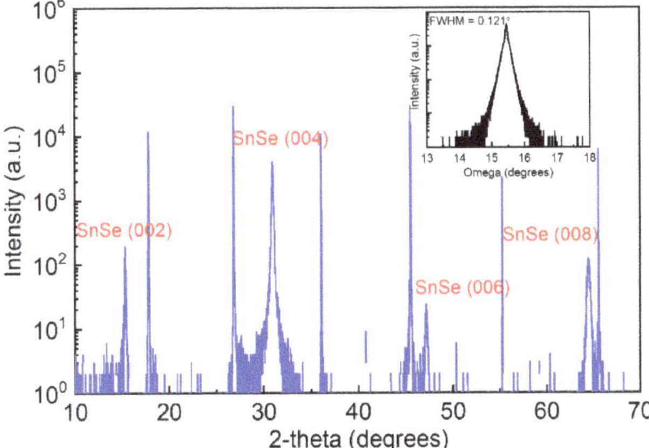

Figure 2. X-ray diffraction θ–2θ scan probing the out-of-plane orientation of the SnSe film on a mica substrate. The inset shows the rocking curve conducted on the SnSe (004) plane.

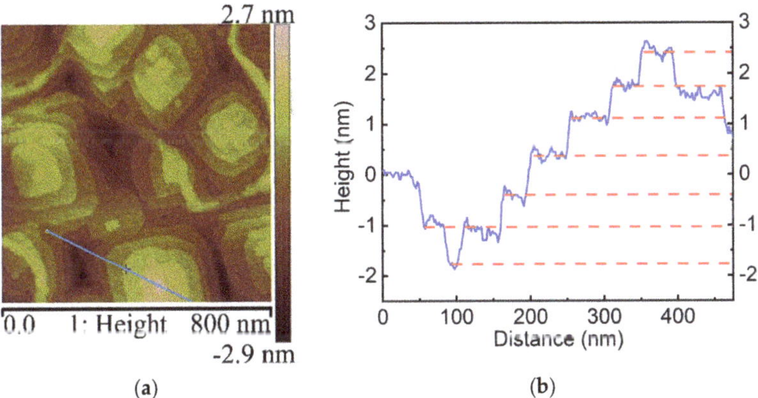

(a)

(b)

Figure 3. (a) AFM morphology of SnSe on a mica substrate. (b) The height profile along the blue line in (a).

First-principle calculations revealed that Sn vacancies in the film played an important role in electrocatalysis acting as reactive sites. Geometry optimization of the SnSe unit cell leading to the abovementioned lattice constants was executed before the simulations. The band structures and density of states (DOS) calculated using the HSE06 method for bulk SnSe are plotted in Figure 4a,b. It can be seen that the band nature of bulk SnSe is indirect and the band gap is computed as 1.20 eV, which agrees with our experimental value of 1.18 eV. To further investigate the electronic structure of the system, the work function of the SnSe (001) surface was simulated as well. The larger the work function, the less likely our system would lose electrons and the more stable it would be. The calculation formula of work function is as follows:

$$\Phi = E_{vac} - E_F \tag{4}$$

where Φ is the electronic work function, E_{vac} is the energy of the vacuum level, and E_F is the energy of the Fermi level. Through calculations, it was found that the energy of the vacuum level was 4.624 eV and the Fermi level was located 0.4563 eV above the valence band maximum. The average electrostatic potential is presented in Figure 4c. The work function of the SnSe (001) surface was thus calculated to be 4.1677 eV, which agrees well with the value of 4.18 eV derived from the UPS measurement as shown in Figure 4d.

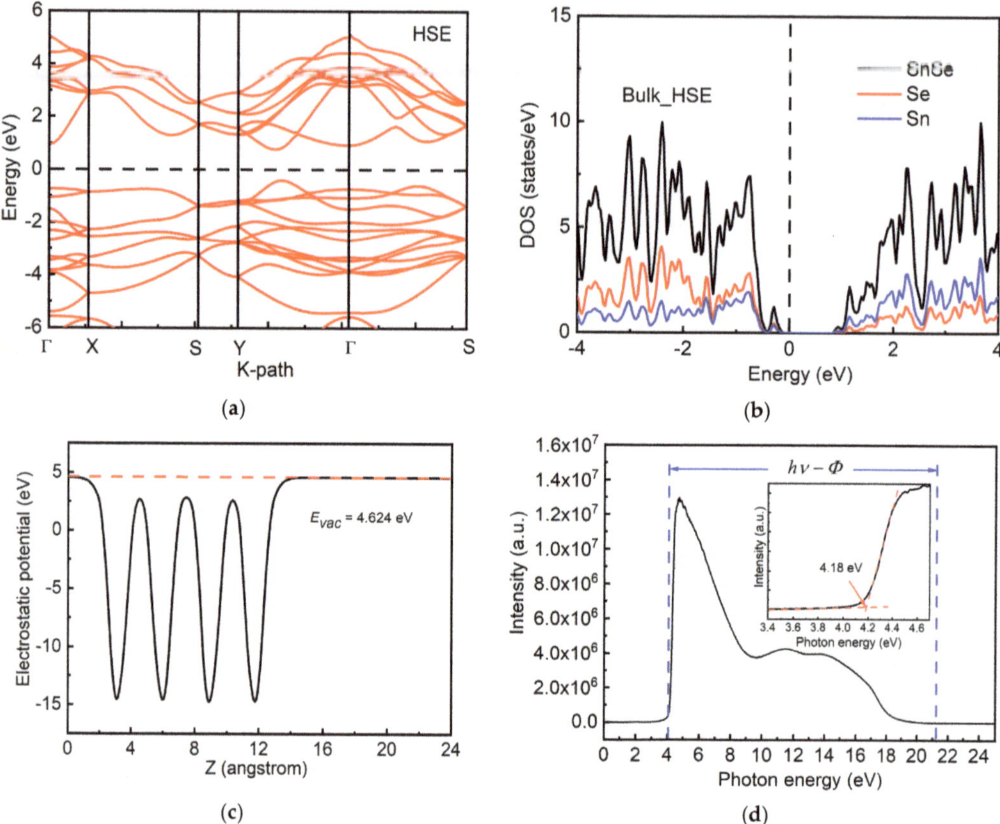

Figure 4. Energy diagrams of SnSe. (**a**) Energy bands in the HSE functional calculations. (**b**) DOS in the HSE functional calculations. (**c**) Average electrostatic potential of the SnSe (001) surface. The red dashed line shows the energy at the vacuum level. (**d**) Work function of SnSe measured by UPS.

HER activity was evaluated by plotting a two-state HER free energy diagram [31], which contains the initial $H^+ + e^-$ state, the intermediate adsorbed H* and the final $\frac{1}{2}H_2$ product. It is well-known that an optimum HER site has a free energy change ΔG_{H*} of hydrogen adsorption close to zero [34,35]. The HER performance of the SnSe monolayer is summarized in Figure 5. It is clearly observed that the clean basal surface of the SnSe monolayer possesses a ΔG_{H*} value of 1.54 eV, demonstrating relatively poor HER activity. However, when an Sn defect site was introduced, it was obviously found that ΔG_{H*} substantially decreased to 0.66 eV, namely by a factor of 2.3, indicating that the HER performance of SnSe can be greatly boosted by introduction of Sn vacancies acting as reactive sites.

To understand the effect of Sn vacancies on the electronic structure of the SnSe monolayer, the density of states was again computed with the presence of an Sn vacancy. Additional electronic states near the Fermi level appeared within the bandgap of SnSe as shown in Figure 6. Thus, the electrical conduction of V_{sn} was substantially enhanced, suggesting that electrons can easily transfer to the reactive sites on the surface, which is beneficial for continuous hydrogen generation. Therefore, DFT calculations demonstrated that Sn vacancies increased the electrical conductivity and reduced the ΔG_{H*} value, leading to a great enhancement in the HER activity.

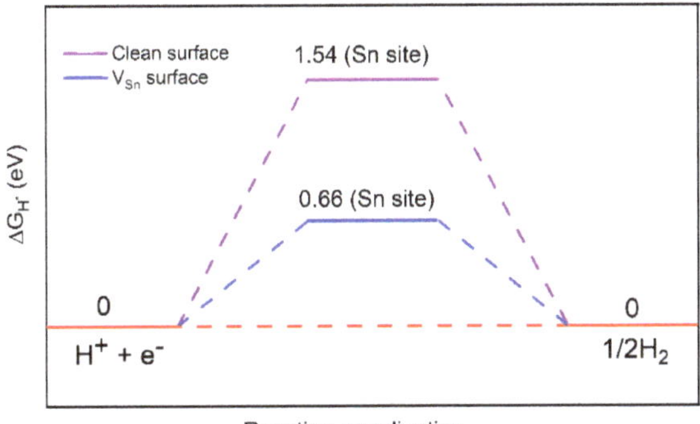

Figure 5. The adsorption Gibbs free energy of H adsorbed on the SnSe monolayer surface. Clean represents the clean SnSe monolayer surface. V_{Sn} represents the SnSe monolayer surface with one Sn defect.

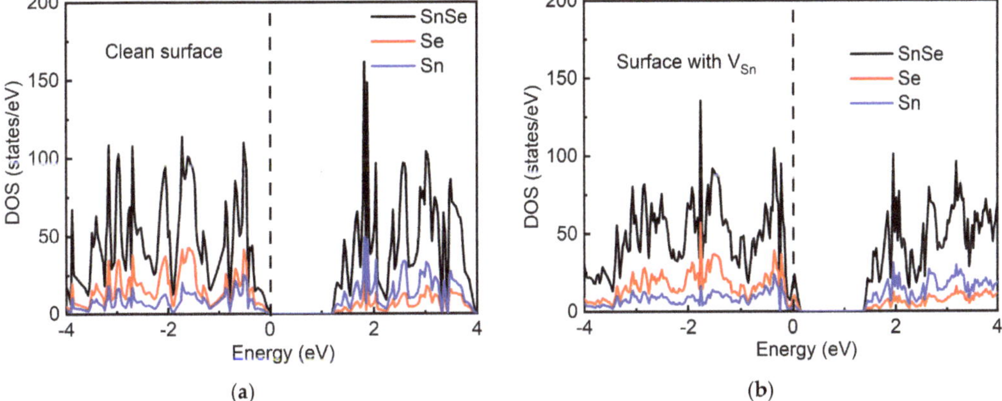

Figure 6. Density of states of two different surfaces. (**a**) Clean SnSe monolayer surface. (**b**) SnSe monolayer surface with one Sn defect.

4. Conclusions

Hydrogen is regarded to be one of the most promising strategies for the development of clean and renewable energy, especially pushed by the carbon neutrality pledges from companies and governments around the world. For clean energy conversion, 2D materials have attracted much attention due to their unique structural and electronic properties, among which SnSe has been recognized as an economical alternative to expensive platinum-based catalysts for hydrogen evolution. In this work, SnSe layers with excellent crystallinity were prepared on a mica substrate using the MBE technique. The films exhibited a perfect layered structure and p-type conductivity, which were attributed to Sn vacancies. However, Sn defects can be easily and accurately regulated by a separate elemental Sn source in a wide range to meet the application requirements. First-principle calculations via the VASP revealed that it is because of the Sn defects more reactive sites are introduced, substantially lowering the Gibbs free energy of H adsorption on the SnSe surface and boosting the HER activity. Enhanced hydrogen evolution performance through controllable defect engineering demonstrated that such 2D SnSe shows great promise for hydrogen generation applications.

Supplementary Materials: The following are available online. Figure S1: XRD in-plane phi scan of SnSe (016) with respect to the SnSe (001) plane; Figure S2: Variation of SnSe sheet resistivity with the tin compensation source temperature.

Author Contributions: Conceptualization, S.W.; methodology, Q.W. and S.W.; software, Q.W. and M.W.; validation, Y.L. (Yong Li) and Y.L. (Yanfang Liu); formal analysis, S.W. and Q.Z.; investigation, Q.Z. and M.W.; resources, S.W.; data curation, S.W.; writing—original draft preparation, Q.W. and S.W.; writing—review and editing, Y.C. and S.W.; supervision, S.W.; project administration, S.W.; funding acquisition, Y.C. and S.W. All authors have read and agreed to the published version of the manuscript.

Funding: This research was funded by the National Natural Science Foundation of China (Nos. 52062045 and 12047575), the Central Government Funds for Local Scientific and Technological Development (No. XZ202101YD0019C), and the Central Government Funds for the Reform and Development of Local Colleges and Universities (No. ZCKJ 2020-11).

Institutional Review Board Statement: Not applicable.

Informed Consent Statement: Not applicable.

Data Availability Statement: Not applicable.

Conflicts of Interest: The authors declare no conflict of interest.

References

1. Muradov, N.Z.; Veziroğlu, T.N. "Green" path from fossil-based to hydrogen economy: An overview of carbon-neutral technologies. *Int. J. Hydrogen Energy* **2008**, *33*, 6804–6839. [CrossRef]
2. Mallapaty, S. How China could be carbon neutral by mid-century. *Nature* **2020**, *586*, 482–483. [CrossRef]
3. Walter, M.G.; Warren, E.L.; McKone, J.R.; Boettcher, S.W.; Mi, Q.; Santori, E.A.; Lewis, N.S. Solar water splitting cells. *Chem. Rev.* **2010**, *110*, 6446–6473. [CrossRef]
4. Qi, J.; Zhang, W.; Cao, R. Solar-to-hydrogen energy conversion based on water splitting. *Adv. Energy Mater.* **2017**, *1701620*, 1–16. [CrossRef]
5. Razek, S.A.; Popeil, M.R.; Wangoh, L.; Rana, J.; Suwandaratne, N.; Andrews, J.L.; Watson, D.F.; Banerjee, S.; Piper, L.F.J. Designing catalysts for water splitting based on electronic structure considerations. *Electron. Struct.* **2020**, *2*, 023001. [CrossRef]
6. Li, X.; Zuo, X.; Jiang, X.; Li, D.; Cui, B.; Liu, D. Enhanced photocatalysis for water splitting in layered tin chalcogenides with high carrier mobility. *Phys. Chem. Chem. Phys.* **2019**, *21*, 7559–7566. [CrossRef] [PubMed]
7. Xu, L.; Yang, L.; Johansson, E.M.J.; Wang, Y.; Jin, P. Photocatalytic activity and mechanism of bisphenol a removal over TiO_2-x/rGO nanocomposite driven by visible light. *Chem. Eng. J.* **2018**, *350*, 1043–1055. [CrossRef]
8. Miyoshi, A.; Nishioka, S.; Maeda, K. Water Splitting on Rutile TiO_2-Based Photocatalysts. *Chem. Eur. J.* **2018**, *24*, 18204–18219. [CrossRef]
9. Wang, Q.; Hisatomi, T.; Ma, S.S.K.; Li, Y.; Domen, K. Core/shell structured La- and Rh-codoped $SrTiO_3$ as a hydrogen evolution photocatalyst in Z-scheme overall water splitting under visible light irradiation. *Chem. Mater.* **2014**, *26*, 4144–4150. [CrossRef]
10. Wang, Q.; Hisatomi, T.; Jia, Q.; Tokudome, H.; Zhong, M.; Wang, C.; Pan, Z.; Takata, T.; Nakabayashi, M.; Shibata, N.; et al. Scalable water splitting on particulate photocatalyst sheets with a solar-to-hydrogen energy conversion efficiency exceeding 1%. *Nat. Mater.* **2016**, *15*, 611–615. [CrossRef]
11. Berglund, S.P.; Flaherty, D.W.; Hahn, N.T.; Bard, A.J.; Mullins, C.B. Photoelectrochemical oxidation of water using nanostructured $BiVO_4$ films. *J. Phys. Chem. C* **2011**, *115*, 3794–3802. [CrossRef]
12. Kudo, A.; Omori, K.; Kato, H. A novel aqueous process for preparation of crystal form-controlled and highly crystalline $BiVO_4$ powder from layered vanadates at room temperature and its photocatalytic and photophysical properties. *J. Am. Chem. Soc.* **1999**, *121*, 11459–11467. [CrossRef]
13. Liu, Y.; Kanhere, P.D.; Wong, C.L.; Tian, Y.; Feng, Y.; Boey, F.; Wu, T.; Chen, H.; White, T.J.; Chen, Z.; et al. Hydrazine-hydrothermal method to synthesize three-dimensional chalcogenide framework for photocatalytic hydrogen generation. *J. Solid State Chem.* **2010**, *183*, 2644–2649. [CrossRef]
14. Nie, L.; Zhang, Q. Recent progress in crystalline metal chalcogenides as efficient photocatalysts for organic pollutant degradation. *Inorg. Chem. Front.* **2017**, *4*, 1953–1962. [CrossRef]
15. Yuan, Y.P.; Ruan, L.W.; Barber, J.; Joachim Loo, S.C.; Xue, C. Hetero-nanostructured suspended photocatalysts for solar-to-fuel conversion. *Energy Environ. Sci.* **2014**, *7*, 3934–3951. [CrossRef]
16. Maeda, K. Z-scheme water splitting using two different semiconductor photocatalysts. *ACS Catal.* **2013**, *3*, 1486–1503. [CrossRef]
17. Sasaki, Y.; Nemoto, H.; Saito, K.; Kudo, A. Solar water splitting using powdered photocatalysts driven by Z-schematic interparticle electron transfer without an electron mediator. *J. Phys. Chem. C* **2009**, *113*, 17536–17542. [CrossRef]
18. Wang, Q.; Li, Y.; Hisatomi, T.; Nakabayashi, M.; Shibata, N.; Kubota, J.; Domen, K. Z-scheme water splitting using particulate semiconductors immobilized onto metal layers for efficient electron relay. *J. Catal.* **2015**, *328*, 308–315. [CrossRef]

19. An, L.; Zang, X.; Ma, L.; Guo, J.; Liu, Q.; Zhang, X. Graphene layer encapsulated $MoNi_4$-$NiMoO_4$ for electrocatalytic water splitting. *Appl. Surf. Sci.* **2020**, *504*, 144390. [CrossRef]
20. Yang, L.; Wang, P.; Yin, J.; Wang, C.; Dong, G.; Wang, Y.; Ho, W. Engineering of reduced graphene oxide on nanosheet–g-C_3N_4/perylene imide heterojunction for enhanced photocatalytic redox performance. *Appl. Catal. B Environ.* **2019**, *250*, 42–51. [CrossRef]
21. Jian, W.; Cheng, X.; Huang, Y.; You, Y.; Zhou, R.; Sun, T.; Xu, J. Arrays of ZnO/MoS_2 nanocables and MoS_2 nanotubes with phase engineering for bifunctional photoelectrochemical and electrochemical water splitting. *Chem. Eng. J.* **2017**, *328*, 474–483. [CrossRef]
22. Najafi, L.; Bellani, S.; Oropesa-Nuñez, R.; Prato, M.; Martín-García, B.; Brescia, R.; Bonaccorso, F. Carbon Nanotube-Supported $MoSe_2$ Holey Flake: Mo_2C Ball Hybrids for Bifunctional pH-Universal Water Splitting. *ACS Nano* **2019**, *13*, 3162–3176. [CrossRef]
23. Wang, Y.; Zhao, S.; Wang, Y.; Laleyan, D.A.; Wu, Y.; Ouyang, B.; Ou, P.; Song, J.; Mi, Z. Wafer-scale synthesis of monolayer WSe_2: A multi-functional photocatalyst for efficient overall pure water splitting. *Nano Energy* **2018**, *51*, 54–60. [CrossRef]
24. Gao, J.; Tay, Q.; Li, P.Z.; Xiong, W.W.; Zhao, Y.; Chen, Z.; Zhang, Q. Surfactant–Thermal method to synthesize a novel two-Dimensional Oxochalcogenide. *Chem. Asian J.* **2014**, *9*, 131–134. [CrossRef]
25. Lu, C.; Zhang, Y.; Zhang, L.; Yin, Q. Preparation and photoelectrochemical properties of SnS/SnSe and SnSe/SnS bilayer structures fabricated via electrodeposition. *Appl. Surf. Sci.* **2019**, *484*, 560–567. [CrossRef]
26. Wang, S.F.; Fong, W.K.; Wang, W.; Surya, C. Growth of highly textured SnS van der Waals epitaxy on mica using a novel SnSe buffer layer. *Thin Solid Film* **2014**, *564*, 206–212. [CrossRef]
27. Wang, S.F.; Wang, W.; Fong, P.W.K.; Yu, Y.; Surya, C. Tin compensation for the SnS based optoelectronic devices. *Sci. Rep.* **2017**, *7*, 39704. [CrossRef] [PubMed]
28. Do, T.N.; Idrees, M.; Amin, B.; Hieu, N.N.; Phuc, H.V.; Hoa, L.T.; Nguyen, C.V. First principles study of structural, optoelectronic and photocatalytic properties of SnS, SnSe monolayers and their van der Waals heterostructure. *Chem. Phys.* **2020**, *539*, 110939. [CrossRef]
29. Wang, Q.; Yu, W.; Fu, X.; Qiao, C.; Xia, C.; Jia, Y. Electronic and magnetic properties of SnSe monolayers doped by Ga, In, As, and Sb: A first-principles study. *Phys. Chem. Chem. Phys.* **2016**, *18*, 8158–8164. [CrossRef] [PubMed]
30. Luo, M.; Xu, Y.; Shen, Y. Magnetic properties of SnSe monolayer doped by transition-metal atoms: A first-principle calculation. *Results Phys.* **2020**, *17*, 103126. [CrossRef]
31. Tang, Y.; Cheng, F.; Li, D.; Deng, S.; Chen, Z.; Sun, L.; Liu, W.; Shen, L.; Deng, S. Contrastive thermoelectric properties of strained SnSe crystals from the first-principles calculations. *Phys. B* **2018**, *539*, 8–13. [CrossRef]
32. Zhang, B.; Fu, X.; Song, L.; Wu, X. Computational screening toward hydrogen evolution reaction by the introduction of point defects at the edges of group IVA monochalcogenides: A first-principles study. *J. Phys. Chem. Lett.* **2020**, *11*, 7664–7671. [CrossRef] [PubMed]
33. Guo, S.; Yuan, L.; Liu, X.; Zhou, W.; Song, X.; Zhang, S. First-principles study of SO_2 sensors based on phosphorene and its isoelectronic counterparts: GeS, GeSe, SnS, SnSe. *Chem. Phys. Lett.* **2017**, *686*, 83–87. [CrossRef]
34. Noerskov, J.K.; Bligaard, T.; Logadottir, A.; Kitchin, J.R.; Chen, J.G.; Pandelov, S.; Stimming, U. Trends in the exchange current for hydrogen evolution. *Phys. Inorg. Chem.* **2005**, *36*, 12154. [CrossRef]
35. Gao, D.; Zhang, J.; Wang, T.; Xiao, W.; Tao, K.; Xue, D.; Ding, J. Metallic Ni_3N nanosheets with exposed active surface sites for efficient hydrogen evolution. *J. Mater. Chem. A* **2016**, *4*, 17363–17369. [CrossRef]

MDPI
St. Alban-Anlage 66
4052 Basel
Switzerland
Tel. +41 61 683 77 34
Fax +41 61 302 89 18
www.mdpi.com

Molecules Editorial Office
E-mail: molecules@mdpi.com
www.mdpi.com/journal/molecules